Five Decades of Research
in Nuclear Science

Five Decades of Research in Nuclear Science

John R. Huizenga

MELIORA PRESS
An Imprint of the University of Rochester Press

First published 2009

Meliora Press is an imprint of the
University of Rochester Press
668 Mt. Hope Avenue, Rochester, NY 14620, USA
www.urpress.com

and Boydell & Brewer, Ltd.
P.O. Box 9, Woodbridge, Suffolk IP12 3DF, UK
www.boydellandbrewer.com

ISBN: 978-1-58046-320-1

Library of Congress Cataloging-in-Publication Data

Huizenga, John R. (John Robert), 1921–
    Five decades of research in nuclear science / John R. Huizenga.
       p. cm.
    ISBN 978-1-58046-320-1 (hardcover : alk. paper) 1. Huizenga, John R. (John Robert), 1921– 2. Nuclear physicists—United States—Biography. 3. Nuclear physics—Research—United States—History. 4. Nuclear physics—Research—New York (State)—Rochester–History. 5. Argonne National Laboratory. 6. University of Rochester. I. Title.
    QC774.H85A3 2009
    539.7092—dc22
    [B]
                                                                          2009031822

British Library Cataloguing-in-Publication Data
A catalogue record for this item is available from the British Library.

Printed in the United States of America.
This publication is printed on acid-free paper

To
My Research Collaborators

Contents

Illustrations

x   *Illustrations*

Table

Preface

In this book I summarize my research in the field of nuclear science carried out at the Argonne National Laboratory and the University of Rochester during the last half of the twentieth century. Also included are descriptions of selected travels (e.g., to the USSR in 1966 and China in 1979) and other activities related to my professional career.

As a recipient of several national awards and honors, including election into the National Academy of Sciences in 1976, I've enjoyed a highly productive and distinguished career in nuclear science spanning half of a century. During this long time period, I interacted with a broad segment of the nuclear science community, co-authoring peer-reviewed research papers with 180 different collaborators, including students, postdoctoral associates, and colleagues, both inside and outside my home institutions. As a result, the research I describe in this manuscript was a joint effort with able assistance from many collaborators. Worthy of special mention were highly productive associations with professors W. U. Schröder and R. Vandenbosch, including co-authorship of books. I wish to acknowledge the dedicated assistance of my collaborators and thank each and every one for his or her contributions.

My decision to write a summary of my research and associated professional activities was made only recently, in 2007, at the age of 86. An influential consideration in my decision process was that such an activity would at least serve as a diversion from my amateurish golf game. In writing this summary I've had to make selective choices, often arbitrary, of which research results to include; however, I have attempted to include a representative sampling of my research in the text and accompanying figures. For those interested in more details, an entire set of reprints of my publications is available in eight large bound volumes. Volumes I to VIII include, respectively, the years 1950–59, 1960–65, 1966–70, 1971–76, 1977–80, 1981–83, 1984–86, and 1987–91. These volumes of research reprints have been donated to the Department of Rare Books, Special Collections and Preservation of the Rush Rhees Library of the University of Rochester. Three other large volumes donated to the University of Rochester Library contain awards, photos, and biographical data. In addition, I've donated to the University of Rochester Library four large volumes containing a small sampling of my correspondence. These latter four volumes cover the time periods 1949–67, 1968–77, 1978–83, and 1984–91. Also available in Rush Rhees Library are nine bound volumes of my Annual Research Reports covering the period of 1967–91.

An energetic curiosity induced me to investigate a wide range of topics in nuclear science, utilizing chemical and physical techniques and performing both off-line and on-line experiments. My research efforts emphasized studies of transuranic nuclei, nuclear fission, nuclear reactions, nuclear level densities, and the cosmic abundances of the chemical elements.

During my initial years at Argonne, I participated in a research program to irradiate ^{239}Pu in high-flux-neutron reactors for long periods of time to produce transplutonium nuclei, including possible new elements. This then largely unexplored region of the periodic table was rightly predicted to be a rich and fertile source of new chemical and nuclear information. The production and study of these new highly deformed actinide nuclei are of basic scientific interest, making possible more extended studies of such fields as nuclear structure, nuclear systematics, nuclear fission, and heavy-element chemistry. In addition, very long irradiations of ^{239}Pu opened up the possibility of making sufficient amounts of transplutonium nuclei to be used as targets in subsequent accelerator experiments, enabling one to probe their excited-state structures.

On November 1, 1952, the first thermonuclear device (named "Mike") was tested on an island in the South Pacific. This event caused high excitement in our research group, recognizing the possibility that debris from "Mike" contained new elements. Our early analyses of elemental plutonium and curium samples from the debris indeed showed that high-mass isotopes of these elements had been produced, confirming that "Mike" produced a gigantic instantaneous number of neutrons. These titillating results spread like wildfire from the Atomic Energy Commission's office in Washington, DC, to scientists in other national laboratories. The race for the discovery of new elements was under way! Research teams from Los Alamos and the Lawrence Berkeley National Laboratory joined in the search. Eventually, scientists from all three of the laboratories collaborated in a joint publication announcing the discovery of the new elements *einsteinium* and *fermium*.

After returning from a sabbatical leave on a Fulbright Fellowship in Amsterdam, I initiated a study of the abundance of selected elements in meteorites, especially chondrites, which are a class of stony meteorites thought to give a good representation of the cosmic element abundances. My interest in this subject was piqued by the observation that the two very different neutron capture processes studied in our labratory each had a very similar astrophysical counterpart. Supernovae explosions, like the "Mike" explosion, release gigantic numbers of neutrons for a small fraction of a second; whereas red giant stars, like reactors, produce a steady flux of neutrons for long periods of time. Each of these processes produces a unique fingerprint of heavy element abundances.

Initially introduced to nuclear fission in early 1944 on the Manhattan Project, I worked for years to better understand this extremely complex, but fascinating, nuclear reaction. A variety of experiments were designed to examine the intermediate transition-state nucleus, the highly deformed saddle-point configuration through which the nucleus passes on the way to fission. Studies of the mass, energy, and angular distributions of the fission fragments were made for a wide range of targets, projectiles, and energies. On another front, major progress was made in understanding the way excited compound nuclei de-excite, with emphasis on the decay competition between fission and other open-exit channels for actinide nuclei. To assist in the interpretation of these experiments, extensive experimental and theoretical studies were conducted of the dependence of the nuclear level density on excitation energy, angular momentum, isospin, nuclear shells, and nuclear deformation. Our multipronged investigations of nuclear fission culminated in a popular reference book entitled *Nuclear Fission*, which I co-authored with Professor R. Vandenbosch, a collaborator at the Argonne National Laboratory.

Soon after spending a year as a Guggenheim Fellow and Visiting Professor at the University of Paris-Orsay, I moved to the University of Rochester as Professor of Chemistry and Physics. The university's Nuclear Structure Research Laboratory had just acquired a state-of-the-art Emperor tandem Van de Graaff. I immediately took advantage of this facility by embarking on a systematic study of the single-particle and collective excited states of actinide nuclei by reaction spectroscopy. From my Argonne days I had access to isotopically enriched targets of exotic actinide nuclei. Nucleon transfer reactions produced single-particle and/or hole states (known as Nilsson states) and inelastic scattering produced predominantly collective states. These direct reaction studies produced a large amount of new data on the spins and energies of neutron, proton, and collective states of highly deformed actinide nuclei and contributed to the rapid development of theoretical models of such nuclei.

The decay modes of actinide muonic atoms were investigated at the Los Alamos Meson Physics Facility (LAMPF). Our experiments, designed to determine the muon capture rates in neutron rich, actinide nuclei, measured the time difference between the arrival of a muon in the target and the appearance of any reaction product associated with the capture of the muon by the nucleus. Of particular interest is the prompt fission resulting from internal conversion, where the transition energy between inner shells of the muonic atom is transferred directly to the nucleus. The muon is not annihilated but can be used to probe the dynamics of fission. The muon was found to attach itself predominately to the heavy fission-fragment atom during the fission process.

A third major research program I carried out at Rochester was a study of heavy-ion-induced reactions. A new type of reaction mechanism was discovered, which became known as strongly damped or deep inelastic collisions (in some respects, the inverse of fission). Detailed studies were made of the energy dissipation, the nucleon transfer, and the microscopic time scale associated with these collisions. This new process yields two fragments in the exit channel, each with Gaussian Z (atomic number) distributions centered near the Z values of the projectile and target. The variances in the Z distributions are strongly correlated with the magnitude of the energy dissipation. Large amounts of radial kinetic energy are dissipated in these reactions, resulting in kinetic energies of the fragments in the exit channel being as small as the Coulomb repulsive energy of two touching highly deformed fragments.

The experimental relations between charge (Z), mass (A), and energy loss of the fragments were analyzed in terms of an independent-particle transport model. In this model, transport of mass, charge, energy, and linear and angular momentum in a strongly damped collision is attributed to the stochastic exchange of individual nucleons between reaction partners. Quantitative agreement between experiment and theory was obtained, including the energy-loss dependence of isobaric and isotopic fragment distributions, when account is taken of the Pauli exclusion principle. In collaboration with Professor W. U. Schröder, a 685-page article entitled "Damped Nuclear Reactions" was published by Plenum Press.

Acknowledgments

I wish to thank Debra Haring for suggesting that I write this manuscript. I especially thank Debra for her special help in preparing this book for publication, as well as her student assistant, Sarah Rudzinskas (Class of 2011), for help in typing the manuscript.

Thanks are also due to Professor W. U. Schröder, John Bertola (Class of 2009), and Sarah Rudzinskas for their help in preparing the figures.

The support of the Department of Chemistry, the Department of Physics and Astronomy, and Professor W. U. Schröder in publishing this volume is gratefully acknowledged.

The research described in this volume was supported by the US Atomic Energy Commission (1949–75), the US Energy and Development Administration (1975–77), and the US Department of Energy (1978–94).

I. Early Years and Time on the Manhattan Project

I was born on April 21, 1921, in Fulton, Illinois, and I spent my early years on livestock and grain farms in northwestern Illinois. At the age of five, I entered a one-room school, having the same teacher for all eight grades. Such an arrangement offered the possibility to listen to the lessons of higher grades. After two years at Erie High School, my family moved to a much larger farm near Morrison. During this Depression period, I worked full time for two years on the farm. Returning to Morrison High School, I was inspired by excellent teachers of mathematics and physics to pursue a scientific career, graduating in 1940 as president of my senior class.

Entering Calvin College in Grand Rapids, Michigan, I majored in mathematics and chemistry, due largely to the influence of chemistry professor John DeVries and by the college's lack of a good physics program at that time. During the summer of 1942, I studied physics at Northern Illinois University in Dekalb, Illinois. At the end of my junior year I entered a written competition solving differential equations and was awarded the Rinck Memorial Prize in Mathematics. The following summer, I spent a full semester at the University of Illinois taking four advanced chemistry courses. By this time, five hours short of a BA degree, I returned to Calvin four weeks after the opening of the fall semester and completed requirements for graduation in December 1943. In January 1944, I enrolled in graduate school in physical chemistry at the University of Illinois; however, pressure from my World War II draft board precipitated my decision to travel to Chicago seeking a commission in the Navy. On return to campus, recruiters for the Manhattan Project convinced me of the importance to utilize my scientific training in an exciting and militarily important secret project. Immediately, I accepted and was ordered to report to a nondescript building in downtown Knoxville, Tennessee. Here, I was indoctrinated on the seriousness of high security and processed for a security clearance. Later bussed to the remote site of Oak Ridge, Tennessee, I was assigned to begin work at Y12, the site where uranium-235 was to be enriched by large magnetic separators known as *calutrons*.

I was placed in a section under the direction of Dr. A. E. Cameron, an expert in mass spectroscopy. This section had the responsibility of making isotopic-analysis measurements of uranium samples following various degrees of enrichment of the uranium-235.

I was first introduced to nuclear fission at this time, a subject in which I later published many research papers and co-authored a popular reference book. My group at Oak Ridge had the task of measuring isotopic abundances of uranium isotopes of samples enriched in uranium-235 by a thermal-neutron fission-counting technique. The uranium was electroplated onto platinum disks, weighed, and fission counted using a large radium-beryllium neutron source in a tank of water to moderate the neutrons to thermal energies. This method of determining the isotopic abundance of the uranium isotopes is successful because uranium-235 has a large thermal neutron fission cross section while uranium-238 does not fission with thermal neutrons.

Soon after arriving in Oak Ridge, the director of the Fission Counting Laboratory appointed me to be laboratory supervisor. This put me in charge of four groups, each with a supervisor and eight employees. Our laboratory had a separate building and operated twenty-four hours a day, seven days a week, consisting of twenty-one shifts per week. Hence, each group worked five eight-hour shifts per week and once every four weeks worked a sixth shift. Fortunately, my position allowed me to work days, although sometimes long ones. It was my responsibility to see that all shifts were fully staffed and performing their share of isotopic analyses in an efficient and timely manner. I arrived at work early enough to touch base with the after-midnight shift and would remain late enough to make contact with the evening shift.

I lived in one of the many dormitories in Oak Ridge, which were hastily built and poorly constructed. Meals were provided in large cafeterias, located in the dormitory area and in the work area. Security was high, especially so when entering a work area. During whatever leisure time I had, I played tennis and softball in the summer and basketball in the winter.

A nuclear bomb, named "Little Boy" and composed of uranium that was enriched in uranium-235 at Oak Ridge, was dropped on Hiroshima by the *Enola Gay*, a Boeing B-29 bomber at 8:15 AM on the morning of August 6, 1945. Three days later, at 11:02 AM on August 9, 1945, a plutonium bomb named "Fat Man" was dropped from a B-29 bomber named *Bockscar* on Nagasaki. A few months after these events, Oak Ridge employees were allowed to leave. Although I was tempted to stay on permanently at Oak Ridge, given my supervisory position with possible rapid advancement, my urge to return to graduate school prevailed. Hence, in late January, 1946, I left Oak Ridge after an exciting two years.

On February 1, 1946, I married Dorothy (Dolly) Koeze and returned to graduate school at the University of Illinois with a graduate fellowship. During the spring and summer semesters, I completed eight graduate courses in chemistry and mathematics. At the urging of John DeVries, I accepted a teaching position at Calvin College in the fall of 1946; however, anxious to complete my graduate work, I returned in June, 1947, to the University of Illinois and joined the research group of Professor Frederick T. Wall who was engaged in theoretical and experimental studies of polyelectrolytes. One of the outstanding unsolved problems in the field at that time was the degree of ionization and mobility of cations in aqueous solutions of polyacrylic acid and sodium hydroxide. Experiments up to that time gave inconsistent and contradictory results. My Oak Ridge experience led me to explore these problems with radioactive tracers, which I introduced to Professor Wall's group. Radioactive sodium-22, utilized in my PhD thesis research, was produced in the cyclotron of Washington University in St. Louis. In the late 1940s, I was inducted into Phi Beta Kappa at the University of Illinois.

My PhD thesis research investigated the electrolytic properties of aqueous solutions of polyacrylic acid and sodium hydroxide by transference and diffusion experiments using radioactive sodium-22. These pioneering experiments produced, for the first time, reproducible and consistent results for the sodium-ion association as a function of neutralization for various concentrations of polyacrylic acid. These results were published in two papers in the *Journal of the American Chemical Society* (1 and 2). (Throughout, the numbers in parentheses refer to articles in my publication list reproduced in the appendix.) The importance of this research can be judged by the large amount of activity that followed in this area over a long time period, and the many citations to these two papers up to the present time. For example, the volume entitled *Polyelectrolyte Solutions* by S. A. Rice and M. Nagasawa, Academic Press (1961) cited these two papers some eighteen times.

Following the completion of my PhD degree, awarded in June 1949, I chose to return to the study of nuclear science. Although I found research on polyelectrolytes stimulating and possible future work on proteins interesting, my positive experience at Oak Ridge compelled me to pursue a research career in the exciting new research areas of transuranic nuclei, nuclear fission, and nuclear reactions. Consequently, I accepted a position at the newly established Argonne National Laboratory, which at that time was located on the campus of the University of Chicago. This enabled me to audit graduate physics courses, including those of two of the great teachers of physics, Enrico Fermi's course in nuclear physics and Gregor Wentzel's course in mathematical physics.

II. Argonne National Laboratory Years (1949–1954): Discovery of Elements 99 and 100

During the initial years at Argonne, my research emphasized nuclear fission and nuclear reactions of transuranic nuclei. Neutron binding energies were studied (3, 7), including measurements of the photoneutron thresholds (4, 6) of uranium-238 and thoruim-232. The aim of these studies was to determine the energy differences among the four radioactive series (4n, 4n+1, 4n+2, and 4n+3), making it possible to determine other binding energies using closed energy cycles. Knowledge of these energies was used to calculate the masses of heavy nuclei (33) and served as a means of refining empirical atomic mass formulas.

Early on, plans were made to irradiate plutonium-239 in reactors at Argonne (the reactor was located at the new laboratory site southwest of Chicago), Hanford (Washington), Chalk River (Ontario, Canada), and Idaho Falls (Idaho), with the objective of producing heavier plutonium and transplutonium nuclei, including new elements. This largely unexplored heavy region of the periodic table was rightly predicted to be a rich and fertile source of new chemical and nuclear information. In addition to their nuclear reactor aspects, the production and study of these new heavy-deformed nuclei are of basic scientific interest, making possible more extended studies of such fields as nuclear structure, nuclear systematics, nuclear fission, and heavy element chemistry. Irradiation of large amounts of plutonium-239, for extensive periods in the high-neutron-flux Materials Testing Reactor (MTR), opened up the possibility of making transplutonium nuclei in sufficient amounts to be used as targets in subsequent accelerator experiments, probing their excited-state structure.

Estimates of the production yields of these new nuclei required information on the thermal-neutron capture and fission cross sections (15, 32). An examination of the thermal-neutron-fission cross sections revealed that nuclei, which undergo fission with thermal neutrons, contain an odd number of neutrons. This is explained by the fact that the resultant even-neutron compound nucleus has greater excitation energy than the odd-neutron compound nuclei formed from thermal-neutron capture on neighboring isotopic even-neutron nuclei.

Since the individual values of the thermal-neutron-fission cross section (σ_f) and the thermal-neutron-capture cross section (σ_γ) are difficult to predict, we suggested (10, 31) that the ratio, σ_f/σ_γ, be correlated with the energy difference, ΔE, between the effective values of the neutron-binding energy (which is the excitation energy of the compound nucleus A formed by thermal neutron capture on nucleus A-1) and the fission threshold of nucleus A. The logarithm of the ratio σ_f/σ_γ varies smoothly with ΔE over six orders of magnitude. This relationship, illustrated in figure 1, follows from the fact that the radiation width

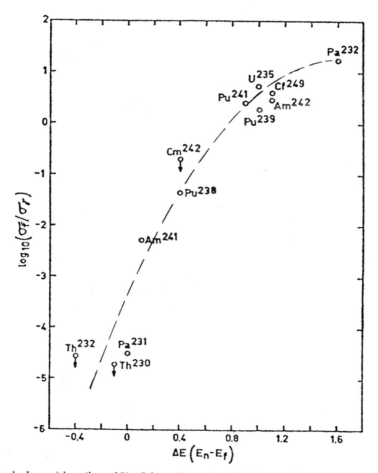

Figure 1. Logarithm (base 10) of the ratio, σ_f/σ_γ, where σ_f and σ_γ are the thermal neutron fission cross section and capture cross section, respectively, is plotted against ΔE (E_n-E_f), where E_n and E_f are the appropriate neutron binding energy and the fission threshold energy, respectively (10).

varies slowly over the excitation energy under consideration, whereas the fission width is highly dependent on ΔE. Hence, if one of the two thermal-neutron cross sections is known, a reliable value of the other, which in some cases may be quite impossible to measure, can be estimated from this correlation.

Among the decay modes of heavy nuclei, decay by spontaneous fission is of particular interest from the nuclear stability point of view. Existing theory in the early 1950s predicted that the spontaneous fission half-lives of nuclei with an even number of both neutrons and protons decrease with increasing values of the liquid drop parameter Z^2/A. Such a relationship predicts that the spontaneous fission half-life of an even-neutron isotope of an even-proton element increases with increasing A. This relationship was fairly successful for the then-known lighter-mass isotopes of several elements. From the limited amount of data available at that time, however, I observed marked discrepancies from this relationship (19). For example, the predicted spontaneous fission half-life of uranium-238 is one-hundred times longer than that for uranium-234. The experimental value is shorter by a factor of two. Nuclei with an excess of neutrons have an additional factor making them unstable to spontaneous fission. Thus, the suggestion was made (19), and later confirmed, that the spontaneous fission lifetimes of even-neutron isotopes of an even-proton element has a maximum at a particular A value, and that as A increases, the lifetime becomes shorter and shorter. This is illustrated in figure 2.

In a further effort to make reliable estimates of spontaneous fission lifetimes of even-even nuclei, the logarithm of the ratio of the spontaneous fission half-life divided by the alpha-particle-emission half-life, R, was plotted versus Z^2/A (26). Such a plot is illustrated in figure 3. It was observed that nearly linear lines (dashed) connect the R values of nuclei differing by two Z units and six A units. For the even-even isotopes of a given element, it is interesting to note that the R values (connected by solid lines) in such a plot increase smoothly with Z^2/A, except for elements fermium and nobelium. One of the most sensational predictions from this plot is that R values become negative for heavy isotopes of some elements. This means that spontaneous fission becomes the principal mode of decay! A remarkable example is californium-254, to be discussed later, where spontaneous fission is indeed the predominant decay mode observed.

Multiple neutron capture on a target of plutonium-239 in a high-flux-neutron reactor proceeds along the beta stability-line. This is illustrated in figure 4 where the solid and open circles represent beta-stable and beta-unstable nuclei, respectively. The value of the atomic mass A increases at constant atomic number Z until the beta-decay rate, Λ_d, begins to compete with the neutron capture rate, Λ_c (the product of the neutron flux and the neutron-capture cross section). At this time, neutron capture begins to occur also in element Z+1. When Λ_d is much larger the Λ_c, most of the neutron capture moves to element Z+1. For example, for neutron irradiation of plutonium-239 in the Materials Testing Reactor in Idaho, neutron capture reaches the first beta unstable nucleus at plutonium-241, which

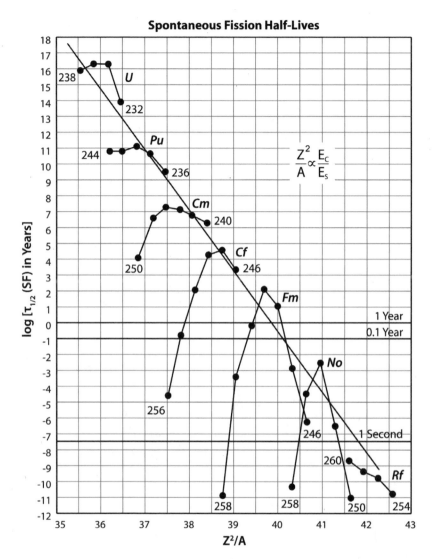

Figure 2. A plot of the logarithm (base 10) of the spontaneous fission half-life versus the parameter Z^2/A, where Z is the atomic number and A is the mass number. The parameter Z^2/A is proportional to the nuclear Coulomb energy divided by the surface energy. Lines connect the values of the spontaneous fission half-lives of the isotopes of each element, showing a maximum and then decreasing half-lives for increasing mass numbers. The sloping solid line illustrates the predicted trend in half-lives with Z^2/A. The three horizontal lines correspond to the half-lives of 1 year, 0.1 year, and 1 second, respectively. This plot, an updated version of such published graphs (19, 31), was prepared for my invited Argonne lecture on June 3, 1996, in celebration of Argonne's 50th anniversary.

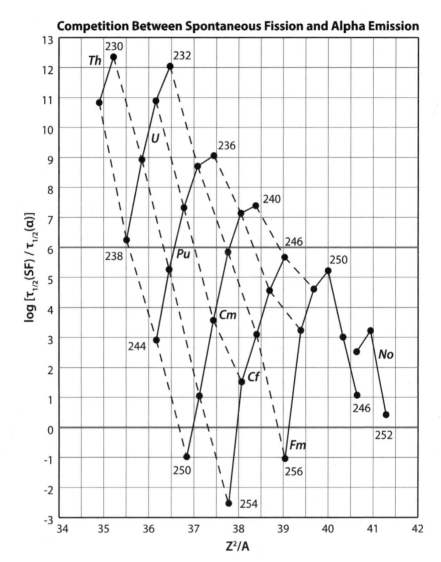

Figure 3. The logarithm (base 10) of the ratio of the spontaneous fission half-life divided by the alpha-particle-decay half-life, R, is plotted against the parameter Z^2/A for isotopes of even-even nuclei. The R values for isotopes of each element are connected by solid lines. The R values of nuclei differing by two units in Z and six units in A are connected by dashed lines. The two horizontal lines represent half-life ratios of 1 and 1 million, respectively. Note that this plot shows that nuclei such as curium-250 and californium-254 decay predominately by spontaneous fission! This plot, an updated version of an earlier similar plot in publication (26), was prepared for my invited Argonne lecture on June 3, 1996, in celebration of Argonne's 50th anniversary.

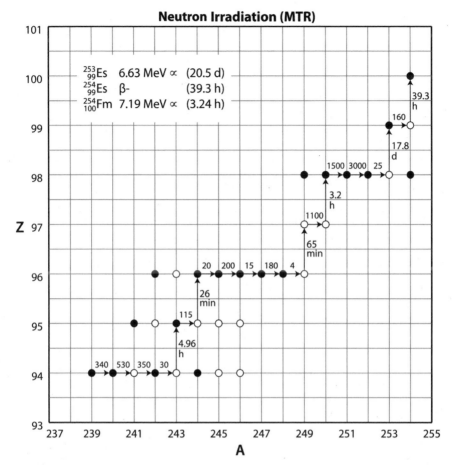

Figure 4. A plot of atomic number Z versus mass number A, illustrating the buildup of transplutonium nuclei in a reactor such as the Materials Testing Reactor in Idaho. The open and solid circles represent beta-unstable and beta-stable nuclei, respectively. The horizontal lines represent neutron capture and the number above the line is the neutron capture cross section in barns (unit of 10^{-24} square centimeters). The vertical lines represent beta decay and the number to the right is the beta-decay half-life. These horizontal and vertical lines represent the dominant buildup pathway for plutonium-239 irradiated in a high-flux neutron reactor. Minor pathways are not included, for example, beta decay at plutonium-241 and berkelium-249. Isotopes of the new elements 99 and 100 are emphasized.

begins to populate americium-241. Because of the 14.4-year half-life of pluto-nium-241, however, neutron capture proceeds predominately to plutonium-243, where $\Lambda_d / (\Lambda_d + \Lambda_c)$ is large and neutron capture moves to americium. In similar fashion, the major neutron-capture path proceeds to curium at A=244, to berke-lium at A=249, to californium at A=250, to element 99 at A=253, and to element 100 at A=254. The predominate neutron-capture path is illustrated in figure 4 by solid-line horizontal arrows. Similar vertical arrows, representing beta decay, show the mass number at which neutron capture has moved to Z+1.

Three large samples (350, 350, and 170 milligrams) of plutonium-239 were irradiated in the core of the Material Testing Reactor, with integrated fluxes of (0.4, 1.1, and 1.4) x 10^{22} neutrons per square centimeter, respectively (32). These very highly radioactive samples were processed at Argonne's new chemis-try hot laboratory (located southwest of Chicago near Lemont, Illinois), which contained a specially equipped hot laboratory. The transplutonium elements were separated from plutonium and fission products by a combination of ion exchange, solvent extraction, and precipitation techniques. The transplutonium elements were then separated from each other by development on a cation exchange resin column with a citrate or glycolate eluting agent at a carefully controlled hydrogen ion concentration.

Although nuclear fission seriously depletes the amount of heavy-element material in the build-up chain at plutonium-239 and 241, and curium-245 and 247, the products of the longer irradiations contained elements from atomic numbers 94 through 100, and nuclei with mass numbers up to 254. The tre-mendous effectiveness of such high-neutron fluxes in making possible very high order neutron-capture reactions is demonstrated by the fact that despite exten-sive reactor neutron irradiations since 1944, nuclei had not previously been built up beyond Z=96 and A=244.

The vast amount of new data on the properties of isotopes of elements 94 to 100 from these high-flux neutron irradiations was published in a series of papers (16, 18, 20, 21, 22, 23, 25, 28, 29, and 38), most of which were published in 1954. A comprehensive summary of these results was reported in a lengthy paper pre-sented at the International Conference on the Peaceful Uses of Atomic Energy held in Geneva, Switzerland, on August 8–20, 1955, and published (32) in the proceedings of this United Nations conference.

The initial publication (16) in the above series, appearing in the *Physical Review* in March 1954, described the nuclear properties of the new elements 99 and 100. This was the first report of these elements produced by pile-irradi-ated plutonium. The separated element 99 fraction contained thousands of counts per minute of 6.6-MeV alpha particles assigned to a nuclide with mass number 253 with a half-life of approximately 20 days. At a low intensity in the element 100 fraction was a 7.2-MeV alpha-particle emitter. Re-irradiation of an element 99 fraction in the Argonne CP-5 reactor substantiated these results. Further re-irradiation of 10,000 alpha counts per minute of another element

99 fraction in the Materials Testing Reactor produced several hundred 7.2-MeV alpha counts per minute of element 100 with mass number 254. The additional neutron irradiation of the element 99 fraction replenished the activity of the beta-decaying isotope of element 99 with mass number 254, which subsequently produces element 100 with the same mass number. The half-lives of this isotope of element 100 with mass number 254 for decay by alpha-particle emission and spontaneous fission were, respectively, 3.3 hours and 220 days. Growth and decay studies of these activities, in an initially separated element 99 fraction, confirmed their genetic relationship with element 99 (A=254). The beta half-life of nuclide (Z=99, A=254) was 37 hours (later shown to be an isomer). These additional results on elements 99 and 100 were published (20) two weeks after the initial publication.

Although the above two articles were early open-literature publications on elements 99 and 100, these two elements had been discovered previously in debris collected from a cataclysmic thermonuclear explosion that occurred on November 1, 1952 (October 31 in the South Pacific). This work will be discussed later as secrecy delayed publication of the discovery paper until 1955.

Returning to the high-flux neutron irradiations of plutonium-239, results of the elemental-plutonium fraction are presented next (16, 18, 29, 32, 38). The residual mass of elemental plutonium from the plutonium-239 sample receiving the largest integrated neutron flux was reduced by a factor of 30. Most unusual is the isotopic composition of this plutonium, as determined by mass spectrometric measurements, containing 98.77 percent plutonium-242 and 0.052 percent plutonium-244, as well as small amounts of lighter isotopes. This sample, highly enriched in plutonim-242, was ideal for studying the decay modes of plutonium-242. Utilizing a combination of alpha-particle and spontaneous-fission activities as well as mass-spectrometric measurements, the alpha-particle and spontaneous-fission half-lives of plutonium-242 were determined to be 3.88×10^5 and 7.06×10^{10} years, respectively (38). The odd-mass plutonium isotopes with mass numbers 243 and 245 are beta emitters with half-lives of 4.98 (14) and 10.1 (32) hours, respectively. Examination of various data from different neutron irradiations of plutonium, yielded neutron capture cross sections of 530 (32), 350 (32), 30 (18), 170 (32), 1.5 (32), and 260 (32) barns (the unit barn has the dimension of 10^{-24} square centimeters) for plutonium isotopes of mass numbers 240, 241, 242, 243, 244, and 245, respectively. The neutron fission cross section of the odd-mass isotope plutonium-241 was measured to be 1100 barns. The ratio of the neutron-fission to neutron-capture cross sections for plutonium-241 is in good agreement with the systematics (10) discussed earlier.

A plutonium-239 sample of 350 milligrams irradiated with an integrated neutron flux of 1.1×10^{22} neutrons yielded approximately 7 milligrams of americium, whose isotopic composition was 99.95 percent americium-243 and 0.05 percent americium-241 (32). Samples of this americium-243 were further irradiated in the Argonne and Materials Testing Reactors to determine whether the

26-minute americium-244 has an electron-capture decay mode (29). If so, this decay mode would be a second production path for plutonium-244. The branching ratio (electron capture/beta decay) was determined to be 0.039 percent, based on the amounts of plutonium-244 and curium-244 collected from the 26-minute americium-244. An enriched sample of plutonium-244 was isolated from neutron-irradiated americium-243, and the spontaneous fission half-life of plutonium-244 was found to be 2.5 x 10^{10} years (29). The measured spontaneous fission half-lives of plutonium-242 and plutonium-244 are consistent with the systematics discussed previously (19, 26).

The mass of the elemental curium fraction, produced by irradiation of a 170-milligram plutonium-239 sample (32) with an integrated neutron flux of 1.4 x10^{22} neutrons per square centimeter, was less than the original plutonium-239 sample mass by a factor of approximately 75. Mass spectrometric analyses (32) of this curium sample showed the isotopic composition of curium isotopes of mass number 245, 246, and 247 to be 1.14, 1.86, and 0.028 percent, respectively.

Analyses of mass spectrometric and alpha-particle-spectral data (21) gave an alpha-particle half-life of 19.2 years for curium-244. The neutron capture cross sections of curium isotopes with mass numbers 244, 245, and 246 calculated from mass spectrometer data were 25, 200, and 15 barns, respectively (21). The neutron fission cross section of curium-245 was measured to be 1800 barns. A value of 115 barns for the neutron capture cross section of americium-243 was calculated from the relative quantities of americium-243 and curium-244 present at the end of the irradiation (21). The alpha-particle half-lives of curium isotopes of mass numbers 244, 245, and 246 were determined by a technique that depends on the growth of the plutonium daughters from a curium sample of known isotopic composition (25). To insure against nonradiogenic plutonium contamination of the curium at the beginning of the first growth period necessitated a second growth period of the plutonium daughters and their separation from curium. The mass composition of the plutonium daughters was measured by a mass spectrometer. Curium-242 and its daughter plutonium-238 served as an internal monitor. The alpha-particle half-lives of the curium isotopes with mass numbers 244, 245, and 246 were calculated (25) to be 18.4, 1.15 x 10^4, and 4.0 x 10^3 years, respectively, based on an alpha-particle half-life of 162.5 days for curium-242.

The elemental berkelium fraction (32), from the same neutron irradiation as the above curium sample, contained some 3 x 10^5 disintegrations per minute of berkelium-249, the daughter of short-lived curium-249, the first beta-decaying isotope in the neutron-capture chain through the curium isotopes. The beta-decay half-life and decay energy of berkelium-249 were measured to be approximately 290 days and 100 keV, respectively (22, 28). In addition, a small alpha-decaying branch (the α/β^- ratio was determined to be approximately 10^{-5}) with the prominent alpha-particle group having an energy of 5.40±0.05 MeV, was observed. The neutron capture cross section of berkelium was determined to be 350 barns and the spontaneous fission half-life was estimated to be greater than 2 x 10^8 years, as expected for an odd-mass isotope.

Elemental californium fractions (labeled I, II, and III) were chemically separated from each of the three earlier mentioned plutonium-239 samples irradiated with integrated fluxes of (0.4, 1.1, and 1.4) x 10^{22} neutrons per square centimeter, respectively (28). Californium samples II and III had masses of approximately (3.5 and 8) x 10^{-7} milligrams, respectively (32). Each of these californium samples contained different mole percentages of the alpha-particle-emitting isotopes with mass numbers 249, 250, 251, and 252. The isotopic composition of sample II was measured by mass spectrometer to be 4.3±0.5, 49±6, 11±3, and 36±5 mole percent, respectively, for the above mass numbers. A small fraction of sample II was left with the earlier described berkelium fraction for five weeks before chemical separation (labeled IIA). Mass spectrometric analyses of this sample gave isotopic abundances of mass numbers 249, 250, 251, and 252 as 28±7, 34±5, 8±1, and 30±3 mole percent, respectively. One notes that californium-249 is significantly enriched. A second fraction of the californium II sample was re-irradiated in the Materials Testing Reactor for six weeks and chemically purified (labeled IIB). The californium-252 in sample IIB was enhanced to greater than 77 mole percent. Californium sample IIC was separated from re-irradiated berkelium, which had then been separated from plutonium irradiated with 1.1 x 10^{22} neutrons per square centimeter. This sample contains only 1.3±0.4 mole percent of californium-252.

The californium alpha-particle energies were determined (28) by measuring the alpha spectra of californium samples II, IIA, IIB, and IIC. The very different isotopic composition of these samples facilitated the assignments of the different alpha-particle energies to specific isotopes of californium. The known alpha-particle energies of curium-242 (6.110 MeV) and bismuth-212 (6.047 MeV) were used as reference standards. The alpha spectra of californium 250 and californium-252 showed prominent alpha-particle groups at 6.033 and 6.117 MeV, respectively, which are emitted in decays to the ground states of the two curium isotopes. The alpha spectra of the above two californium isotopes exhibited characteristic even-even nuclide fine structure peaks. Their energies were 5.99 and 6.08 MeV, respectively, as determined by alpha-particle coincidence measurements with curium L X-rays. Contrary to alpha-particle-energy systematics for transuranium nuclei, the alpha energy for californium-252 is larger than that for californium-250 indicating an irregularity in the nuclear energy surface. Californium sample IIA also contained a small alpha-particle peak at 5.81 MeV. Since a peak of this energy was observed to grow into a beta-emitting berkelium-249 sample, which had previously been separated from all detectable californium alpha activity, the 5.81-MeV alpha group was assigned to be the prominent alpha group of californium-249.

In this pioneering work on the isotopes of californium, preliminary α-particle decay half-lives of californium isotopes with mass number 249, 250, and 252 were determined to be 400±100 years (23), 10.0±2.4 years (28) and 2.2±0.2 years (28), respectively. The (α/sf) disintegration ratios of californium 250 and 252 were measured to be 1460±350 and 30±1 (28). Hence, the spontaneous fission half-lives of these even-even isotopes support the earlier published systematics of such half-lives (19, 26). The half-life of the beta-emitting californium-253 was determined to be 18±3 days by measuring the growth and decay of the 6.61-MeV alpha particles of nuclide (Z=99, A=253) in a chemically separated californium sample as a function of time (28).

During the time period of the high-flux neutron irradiations of plutonium in the Materials Testing Reactor, an unexpected event occurred that markedly increased the intensity of research activity of our Argonne heavy-element group. On November 1, 1952, the first thermonuclear device (named "Mike") was tested by the Los Alamos Scientific Laboratory at Elugelab, one of forty named islands comprising Enewetak Atoll off the Marshall Islands in the South Pacific. The device used liquid deuterium and required cryogenic cooling equipment. Packed with a considerable amount of uranium, it weighed more than 160,000 pounds. The test completely vaporized the island of Elugelab as well as portions of adjacent islands, Sanil and Teiter, leaving an underwater crater more than 160 feet deep and 6,000 feet wide.

In the "Mike" explosion, the uranium was subjected to a pulse of 4±2 moles of neutrons (learned later) delivered in a small fraction of a second. Debris from this explosion was collected by airplanes flying through the "Mike" cloud carrying filters attached to their wings. This was a standard technique used previously by the weapons laboratories to evaluate atmospheric tests. Additional fall-out material was collected from the surface of a neighboring atoll. Tons of coral were processed using bismuth phosphate as a carrier for actinide elements. The pilot plant operation was named "Pay dirt."

The first indication that "Mike" produced a gigantic instantaneous pulse of neutrons was the discovery by our Argonne group of an unusually large mole percentage of plutonium-244 in a chemically separated plutonium sample from the debris. The firm establishment of the substantial enrichment of plutonium-244 was made by a mass spectrometer measurement. This titillating result spread like wildfire from the Atomic Energy Commission (AEC) office in Washington, DC, to the other National Laboratories. The message was loud and clear and stimulated a search for new elements beyond californium, the element with the highest known atomic number at that time of 98. Soon a group at the Los Alamos National Laboratory identified the 10.8-day beta-decaying isotope of plutonium with mass number 246. The race was on. In addition to Argonne and Los Alamos, a group at the Lawrence Berkeley Laboratory joined in the search for transcalifornium elements.

Another important early measurement carried out by our Argonne group was an analysis of the "Mike" curium. The isotopic composition was measured by mass spectrometer to be 70.1±0.6, 27.0±0.6, 2.2±0.1 and 0.7±0.07 percent, respectively, for isotopes of mass number 245, 246, 247, and 248. The yields of the curium isotopes from the "Mike" device display an even-odd mass effect. In going from an odd-neutron isotope to an even-neutron isotope, the yield drops by a factor of approximately 2.9. Going from an even-neutron isotope to an odd-neutron isotope, the yield drops by a factor of approximately 12. Increasing the mass number of a curium isotope by two units decreases the yield by a factor of approximately 35. This decrease in yield by a factor of 35 for every two units of mass increase served as a recipe to estimate the yields of still heavier nuclei produced in the "Mike" device. The prospects for having produced transcalifornium elements appeared to be excellent!

In addition, I was personally excited about the possibility of discovering in the debris new even-neutron-rich isotopes of curium and californium. The "Mike" device offered a unique opportunity to search for such nuclei, and if found, to test further my new ideas on spontaneous fission systematics. The nuclei californium-254 and curium-250 will be discussed later.

Before proceeding to discuss the discovery of elements 99 and 100, I will describe the buildup of heavy nuclei by neutron capture in the "Mike" explosion. The buildup pathway is completely different from that described earlier for a high-flux neutron reactor, where the buildup follows the beta-stability line. In "Mike," neutron capture occurs in only a long-series of uranium isotopes and subsequently these isotopes of uranium decay to more stable nuclei as illustrated in figure 5. The solid and open circles are beta-stable and beta-unstable nuclei, respectively. The solid-horizontal arrows represent neutron capture. The dashed-vertical arrows represent beta decay. Beta decay is a slow nuclear process relative to the short-time period of the neutron pulse associated with a thermonuclear explosion. Hence, all the beta-decay activity occurs after the neutron capture is complete. In "Mike," nuclei were observed with mass numbers up to A=255. The initial reaction process was multiple-neutron capture producing a long series of heavy uranium isotopes, extending at least to uranium-255 (17th order neutron capture). Whether or not this cutoff in mass number is simply a result of the yield recipe for "Mike" described above is an open question. Spontaneous fission systematics suggest that this process may begin to compete with beta decay for such neutron-rich nuclei as uranium-256.

After the arrival of the "Mike" debris in our laboratory, my colleagues and I worked frantically to extract as much scientific information as possible from this once in a lifetime opportunity. We were not only competing with scientific groups at other National Laboratories but also racing against the natural radioactive decay lifetimes of our samples.

By the middle of December 1952, my colleagues and I had confidently assigned a new 6.6-MeV alpha-particle group to element 99. The 6.6-MeV alpha particles eluded in a well-separated peak, ahead of californium (established by a californium-246 tracer) from a Dowex-50 resin column utilizing an ammonium citrate solution at 87° C.

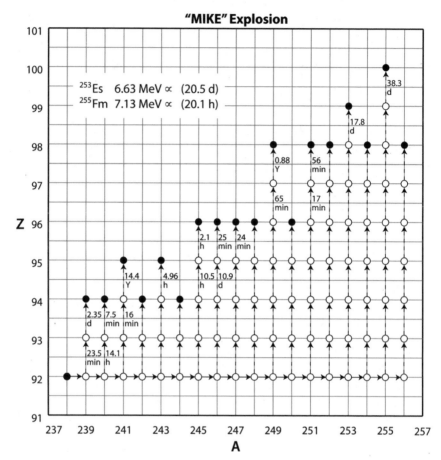

Figure 5. A plot of atomic number Z versus mass number A, illustrating the buildup of transuranium nuclei in the "Mike" thermonuclear explosion. The open and solid circles represent beta-unstable and beta-stable nuclei, respectively. The solid horizontal lines represent neutron capture on uranium isotopes, known to have occurred up to uranium-255. The dashed vertical lines represent the sequential beta decay processes that occur after the neutron capture is completed. The new isotopes of element 99 and 100 are emphasized.

Some controversy developed over this assignment. The Berkeley group assigned this 6.6-MeV alpha group to element 100 with mass number 254. Their assignment was based on their observation of spontaneous fission activity. The Berkeley group's element-99 fraction was contaminated with a trace of californium. Hence, their element-99 fraction displayed considerable spontaneous fission activity, which was certainly not due to an isotope of element 99. Adhering to Z^2/A fission spontaneous fission systematics and not anticipating, as we had, a short spontaneous fission half-life for high-mass californium-254, the Berkeley group mistakenly assigned the 6.6-MeV alpha emitter to element 100. After some discussion, however, the Berkeley group soon agreed that the 6.6-MeV alpha-particles were associated with element 99 with mass number 253. The half-life of nuclide (Z=99, A=253) was measured to be approximately 20 days and is the daughter of 20-day californium-253. More accurate values for these quantities were measured from Materials Testing Reactor data reported earlier in this book.

Some short time after the discovery of element 99, groups at Argonne, Berkeley, and Los Alamos observed a new 7.1-MeV alpha-particle activity. The first two laboratories showed this activity to elude ahead of the 6.6-MeV alpha-emitting isotope of element 99 from a Dowex-50 resin column as mentioned earlier. The 7.1-MeV alpha-particles were assigned to element 100 with mass number 255. These new and exciting results remained unpublished for a couple of years due to security restrictions.

When security was lifted, the involved groups at the three laboratories (Argonne, Berkeley, and Los Alamos) decided amicably to publish the discovery of elements 99 and 100 jointly (30). The authors suggested for the element with atomic number 99 the name einsteinium, after Albert Einstein, and for the element with atomic number 100 the name fermium, after Enrico Fermi. These two world-renowned scientists had died on April 18, 1955, and November 28, 1954, respectively, just prior to submission of our paper.

The Argonne National Laboratory invited me to present a colloquium on June 3, 1996, in celebration of their fiftieth anniversary. It was an honor for me to fulfill this request. I was assigned the topic "The Discovery of Elements 99 and 100: Prelude to a Rich History of Accomplishment in Nuclear Chemistry at Argonne." This lecture gave me the opportunity to review our discovery experiments, as well as to review some of the other important contributions made by the other Argonne nuclear chemists.

During my initial years at Argonne, I developed a strong interest in many aspects of nuclear fission. The subjects of spontaneous fission and thermal-neutron fission have been discussed already. A study of fissionability was initiated by examining the decay probability of moderately excited nuclei by fission and neutron emission. The photofission and photoneutron emission in uranium-238 was studied in an off-line experiment (13). The number of fissions occurring was determined by measuring the number of atoms of molybdenum-99, barium-139 and -140 that were produced during the irradiation. The photoneutron yield was determined by measuring the number of uranium-237 atoms produced. The photoneutron to photofission cross-section ratio in an excitation energy range of 8–10 MeV was measured to be approximately 4.

In another photofission experiment (24), the relative photofission yields of thorium-232, uranium-238, -236, -235, -234, -233, neptunium-237, and plutonium-239 were measured at three betatron energies. The relative yields (assuming uranium-238 is 1.00) in the order of the above list are 0.31, 1.00, 1.43, 2.40,

1.82, 2.54, 2.40, and 3.17, respectively. These relative fissionabilities are correlated with the parameter Z^2/A and show a general increase in fissionability with
this parameter.

Other investigations carried out during my initial years at Argonne included
studies of oxidation-reduction reactions of different valence states of neptunium
(5, 12) and studies of the nuclear radiations of uranium-237 (11), plutonium-
243 (14) and lead-205 (27).

III. Fulbright Year (1954–1955)

My hastily prepared application for a Fulbright Fellowship was successful! My hosts were the Physics Department of the Free University of Amsterdam and the Institute for Nuclear Research. This institute, located on the outskirts of Amsterdam, housed the Philips Cyclotron, the only such machine at an elevation below sea level. The Fulbright Award came as particularly good news as the preceding couple of years had been a very intense period. The Argonne National Laboratory had the enlightened policy, and I assume still has, of augmenting the salary of anyone awarded a national fellowship.

Designated as the one to write the papers on the results from the MTR neutron irradiations, eight such papers were written and published in 1954. In addition, four additional papers on other topics were published in 1954. All of these papers were submitted before the September 1 deadline, the date I was to begin my vacation to prepare for departure. Seven of these papers were published in *Physical Review Letters* and were of short length. However, as the premier journal of the American Physical Society, the threshold for acceptance in *Physical Review Letters* is high, and although short in length, these papers required additional time and special care in preparation.

We sold our home in a southern suburb of Chicago and moved all our household belongings to a rented house near the laboratory. Our long-range plan was to build a new home nearer the laboratory on return. We were successful in subletting our newly rented house to a family beginning work at Argonne. During the first few days of September, we packed up clothing, supplies, and numerous items needed for a family spending a year abroad. My wife and I had four young children at this time; hence, our extensive luggage required a roof rack on our car. All packed on September 5, the six of us departed for New York City. On September 8, we sailed with our car on Holland-American Line's *Maasdam* for the Netherlands, arriving in Rotterdam on September 17, 1954.

We were met by the head of the physics department of the Free University and his English-speaking daughter. The university had rented a house for us in the small city of Laren, a short distance from Amsterdam, and the home of many Dutch artists and painters. In fact, our rented home belonged to a Dutch painter who was spending the winter months painting scenery in Italy and the summer months in Sweden. The rental agreement included a cleaning woman, a long-time employee, who came in one day a week to do the heavy cleaning. In addition, we were able to hire full-time a young woman trained to be a household assistant by a special Dutch school. She was invaluable, arriving after breakfast and staying through dinner in the evening. She took charge of shopping, meal preparation, light household duties, and childcare. She was available also to stay overnight when my wife and I were on a trip.

Our oldest daughter, Linda, was enrolled in first grade in a school nearby on our street. She had completed first grade in the United States the previous year at the age of five, skipping kindergarten. Linda mastered the Dutch language quickly and by Christmas was the star performer in a long public program. Jann and Robert were enrolled in preschool programs, which Dutch children attended at an early age. Although my wife and I were of Dutch ancestry, all of our grandparents having emigrated from the Netherlands, we had not acquired skills in the Dutch language. In our youth it was a period in history when children shunned the opportunity to learn the language of their parents. My progress in Dutch was minimal, all my colleagues insisting they communicate with me by practicing their excellent skills in English. My wife, more talented in language, had much better success in speaking with merchants and others she met on a daily basis in the village.

There was public transportation between Laren and Amsterdam, although I drove some of the time. My Fulbright obligations at the Free University were fulfilled by a series of ten lectures entitled, "Nuclear Properties of Heavy Elements," a subject requiring minimal preparation for me. These lectures were delivered between October 25, 1954, and March 21, 1955. Time was spent also with graduate students doing radioactive decay scheme studies. I had the opportunity to serve on one student's public doctoral examination, a very formal affair requiring formal attire.

A considerable fraction of my time in Amsterdam was spent in study and library research. The previous years had allowed little time for these important activities. Of special interest to me was the research program at the cyclotron. The Argonne Chemistry Division had acquired a new 60-inch cyclotron, and I was making plans to initiate a new research program utilizing this excellent facility. Performing some of the experiments in nuclear fission and nuclear reactions that I envisioned doing in the future required adding on-line techniques to my then-existing arsenal of experimental tools. Plans were made to implement this decision on return to Argonne. Another important aspect of my stay in Amsterdam was the opportunity to visit European laboratories. I made visits and presented colloquia entitled "Production and Nuclear Properties of Elements 97, 98, 99, and 100" at the following institutions:

November 4, 1954	University of Heidelberg, Heidelberg, Germany
November 5, 1954	Max Planck Institute, Mainz, Germany
January 27, 1955	United Kingdom Atomic Energy Research Establishment, Harwell, England
February 12, 1955	Meeting of the Netherlands Physical Society
March 3, 1955	University of Birmingham, Birmingham, England
March 4, 1955	Oxford University, Oxford, England
June 2, 1953	University of Utrecht, Utrecht Netherlands

Colloquia entitled "The Nuclear Fission Process" were presented at the following institutions:

| March 16, 1955 | Municipal University of Amsterdam, The Netherlands |
| June 22, 1955 | Nobel Institute of Physics, Stockholm, Sweden |

I had a particularly interesting experience in Mainz. Professor F. A. Paneth, a distinguished radiochemist and my host at the Max Planck Institute, had not spoken to Professor Fritz Strassman of the University of Mainz, Nobel Laureate and co-discoverer with Otto Hahn of nuclear fission, for some years. My colloquium at the Max Planck Institute enticed Strassman to come over to the institute. Therefore, I was instrumental in re-uniting these two scientific giants. The story was told to me that whenever Strassman sent a sample to the Institute for bombardment, Institute people would interfere by placing absorbers between the accelerator beam and Strassman's sample. I've always been curious to know whether Paneth's and Strassman's relationship improved after my visit to Mainz. Another interesting experience, following my colloquium at Oxford, was participating in the mystical ceremony of dinner at high table at one of Oxford's colleges.

An unfortunate accident occurred during my first sabbatical. On a vacation trip in the early spring to Paris, we had a quirky car accident on a small street near the Sacre Coeur in Montmartre. Driving over a street divider, a circular metallic device some four inches high, the top burr of the divider just caught the underside of our car's transmission. Having driven over many of those in Paris without incident, this particular accident was caused by a sharp dip in the road and the coincident downward spring of the car. Although repair was done in Paris, parts had to be shipped from the United States and considerable time elapsed before repair was completed. On a positive note, the repairs required a couple of extra trips to Paris!

Scientific seminars in Europe, particularly in the Netherlands in the mid-1950s, were held in non-ventilated smoke-filled rooms. After the coffee period, everyone lit up a cigar or pipe and the dense cloud of smoke made it difficult sometimes to see the speaker. It was especially difficult for a nonsmoker, like me, to breathe. I remember, for example, a seminar by Wolfgang Pauli, awarded the Nobel Prize for the discovery of the Exclusion Principle, in which he strutted across the front of the room at a rapid pace from one side to the other as he spoke. Sitting in the front row, it required one to move one's feet on a moment's notice to avoid being stepped on.

During my Fulbright year, I prepared a paper (33) on the atomic masses of nuclei with mass numbers greater than 200, utilizing decay energies for nuclei produced in the MTR irradiations. This effort complimented similar work on lighter nuclei being carried out by A. H. Wapstra at the Institute for Nuclear Research in Amsterdam. Another paper entitled "The Nuclear Fission Process" was prepared and presented at the International Conference on the Peaceful Uses of Atomic Energy held in Geneva on August 8–20, 1955. This contribution was later published in the proceedings of the conference (31). After the Geneva Conference, I returned to Amsterdam. We packed up our luggage, drove to Rotterdam, and sailed for New York on Holland-American Line's *Ryndam*. After unloading our car and belongings from the ship, we made the trip home by car arriving by the end of August. We returned to our rented home and began plans for building a new house in Western Springs, a short distance from Argonne. Within nine months, our new home, which I contracted, was ready for occupancy.

IV. Argonne National Laboratory Years (1955–1964)

Soon after returning from my Fulbright year, I initiated a research program to study the abundance of selected elements in meteorites, especially chondrites, a class of stony meteorites thought to give a good representation of the cosmic elemental abundances. Urey (Phys. Rev. 88 248 [1952]) first suggested that chondrites constitute a valid average composition of the nonvolatile part of solar matter.

My interest was kindled by the publication of Suess's and Urey's classical paper on elemental abundances (Rev. Mod. Phys, 20 53 [1956]) and the speculation about the similar half-lives of a special type of supernova and californium-254. Baade, Hoyle, Burbridge, Burbridge, and Fowler suggested in a paper (Phys. Rev, 103 1145 [1956]) that the exponential decay of these supernovae, between 100 and 640 nights, was due to the decay of californium-254 produced in a supernova explosion (some 10^{47} ergs under the exponential decay).

During the time of our "Mike" and Materials Testing Reactor "MTR" experiments, Fowler and Hoyle spent time in our laboratory, as well as presenting colloquia. Based on the two very different types of neutron capture that I have previously described, Burbridge, Burbridge, Fowler, and Hoyle (BBFH) in a classical paper (Rev. Mod. Phys. 29 547 [1957]) made an analogy between neutron capture in a supernova explosion and the "Mike" explosion, both releasing large pulses of neutrons for a very short period of time. In addition, they made a second analogy between neutron capture in red giant stars and in a neutron reactor, such as the MTR, where both have moderate neutron fluxes for long periods of time, and mass buildup by neutron capture occurs along the beta-stability line. BBFH coined the terms "r process" and "s process," representing neutron capture on a rapid- and slow-time scale, respectively. In the "r process," neutron capture occurs prior to beta decay as discussed earlier in the "Mike" event (see figure 5). In the "s process," beta decay competes with neutron capture leading to the buildup of nuclei along the beta-stability line, as in the MTR experiments described previously (see figure 4). Hence, the astrophysical "r" and "s" neutron-capture processes had their origins in our terrestrial experiments, where

"Mike" demonstrated 17th order neutron capture on uranium-238 in a fraction of a second, and the "MTR" experiments, over a long time period, produced massive neutron capture along the beta-stability line.

It is thought that the "r process" and the "s process" contribute about equally to the nucleosynthesis of heavy nuclei. The heavy elements produced by neutron capture, however, are rare in abundance compared with the abundances of light elements. Some heavy elements are thought to have originated almost exclusively from either the "r" or "s" process, while others are from a combination of both processes. The elemental abundance plotted as a function of mass number shows pronounced structure that has its origin in these two very different neutron capture processes. These peaks in the elemental abundances are associated with magic neutron and proton shells, which have reduced neutron cross sections. For example, for the 126-neutron shell, the "r process" peak is offset to a lower atomic number than the corresponding "s process" peak. This is due to the fact that neutron capture in the "r process" travels below the beta-stability line (see figure 5) and reaches the 126-neutron shell at a lower atomic number.

A. Cosmic Abundances of Selected Elements

Sparked by our results from the "Mike" and "MTR" experiments, which demonstrated "rapid" and "slow" neutron capture, respectively, and the fascinating suggestion that the exponential decay of some supernovae was due to californium-254, I was inspired to initiate experiments in cosmochemistry. In our initial experiments, we determined the thorium content of stone (44, 54) and iron (49) meteorites. Thorium is one of the two heaviest elements found in nature (the other is uranium) and is made exclusively by the "r process." The analyses of several elements in meteorites were made by the neutron-activation technique, employing either the Argonne CP-5 heavy-water reactor or the Material Testing Reactor in Idaho as the source of neutrons.

Eight analyses of five different chondrites gave an average of 3.96×10^{-8} grams of thorium-232 per gram of meteorite. In converting this weight result to an atomic cosmic abundance, we followed Urey's suggestion that silicon has an atomic cosmic abundance of 10^6. In addition, the average silicon content is 0.185 grams of silicon per gram of chondrite, considered to be primitive solar nonvolatile material. Hence, the atomic cosmic abundance of thorium is 0.026, $[(3.96 \times 10^{-8}/232) / (0.185/28.1)]\, 10^6 = 0.026$.

Since the determination of the thorium abundance involves the absolute determination of the neutron flux, capture cross section, absolute counter calibration, and so forth, it was found in practice much easier and more accurate to compare the activity of the unknown sample with that of a known amount of thorium irradiated under the same conditions (known as the flux monitor) and to calculate the amount of thorium present in the sample from its proportionate activity after neutron radiation. By counting the flux monitor under the same geometry that the sample is counted, the counter efficiency need not be determined. Other variables, not amenable to accurate experimental evaluation, are also conveniently eliminated by use of a flux monitor.

The concentration of thorium in iron meteorites was found to be much lower than that in chondrites. The chemical form of the thorium in metal meteorites is uncertain. However, in a private communication from Urey, the thorium in metal meteorites may be dissolved in the iron phase as ThO, ThS, and so on, due to exposure of the metal to high temperature. The thorium concentration in two iron meteorites was found to be in the range of $(0.6–2.0)$ x 10^{-11} gram thorium per gram of meteorite (49). The variability is probably accounted for by nonhomogeneous meteorite samples. In order to obtain greater sensitivity these samples were irradiated in the Materials Testing Reactor.

Our values of the thorium concentration in iron meteorites are lower by at least two orders of magnitude from values determined previously. At these low concentrations the neutron activation technique is truly superior.

During the week of July 7–11, 1957, a small group of people, interested in elemental abundances and nucleosynthesis, held a workshop on "Stellar Evolution and the Abundances of the Elements" at the University of Michigan in Ann Arbor. Included in the attendance were H. C. Urey, F. Hoyle, W. A. Fowler, A. G. W. Cameron, and L. H. Aller. Aller, an experimental astronomer on the faculty of the University of Michigan, was active in determining the abundance of the elements in the sun and main-sequence stars. I was delighted to have been invited to this workshop and to be able to participate in the discussions on elemental abundances and nucleosynthesis with this small group of experts. I had submitted a paper (43) on May 1, 1957, to the *Physical Review* on the subject of "Spontaneous-Fission Half-Lives of Californium-254 and Curium-250." At this time in history there was a strong belief by some astrophysicists that the exponential decay of some supernovae with a 55-night half-life was due to californium-254. As can be seen in figure 5, the "r process" produces californium isotopes with mass numbers 249, 251, 252, 253, and 254. The beta-stable californium-250 is missing because this mass number is held up at curium-250.

At the Ann Arbor workshop, I presented the latest information about the spontaneous-fission half-life of californium-254. Curium-250 and californium-254 are special nuclei insofar as they decay predominantly by spontaneous fission. We had predicted this from an earlier version of figure 3. A sample of californium from the "Mike" event was periodically counted for its spontaneous-fission activity in our laboratory over a period of four years. Since mass number 250 is held up at curium-250, only californium-252 and californium-254 contribute to the spontaneous-fission activity. After four years, more than 25 half-lives of californium-254, we assumed the remaining spontaneous-fission activity was entirely due to californium-252. Assuming a half-life of 2.2 years for californium-252, corrections were made to the earlier measurements of spontaneous-fission activity and a half-life of 56 days was determined for californium-254 (43).

After an initial maximum, the light curve for the supernova in IC 4182 is reported to decay with a 55-night half-life between the period of 100 to 640 nights. If the exponential decline in the light curve, over this long period of time, is due to the decay of californium-254, I found it difficult to account for the absence of a tail in the light curve, near the end of this time period, from californium-252. The reported pure exponential decay for the supernova in IC 4182 implies that the californium-254/californium-252 mole ratio was greater than ten. Possible explanations for production of such a large ratio in supernova are explored in publication (43), however, none are convincing.

The spontaneous-fission half-life of curium-250 was estimated from a curium sample separated from "Mike" debris. The mole percentages of the isotopes in this curium sample were reported previously. The mole percent of curium-250 was too small to determine from a mass spectrometric analysis. However, the mole percent of curium-250 was estimated by interpolating between the "Mike" yields of curium-248 and californium-252 to be 0.02 mole percent. The specific spontaneous-fission activity of the "Mike" curium exceeded the value calculated for the even-mass isotopes of curium-246 and curium-248. Since the odd-mass isotopes do not make a measurable contribution and heavier-even-mass isotopes are ruled out on stability arguments, it is reasonable to assign the additional spontaneous-fission activity to curium-250. Utilizing a curium-246/curium-250 atom ratio of 27.0/0.02 gives a spontaneous-fission half-life of curium-250 of 2×10^4 years, with considerable error.

In publication (43) I speculated, in analogy to astrophysicists' explanation of the enormous energy release during the decay of the supernova in IC 4182 as due to californium-254, that the spontaneous-fission decay of curium-250 may be an important source of energy in old remnants of supernova explosions, for example, in Crab Nebula.

Although astrophysicists' direct association of spontaneous-fission activity, from r-process-produced neutron-rich heavy nuclei, with the exponential decline in the light curves of supernova, was popular in the 1950s, to my knowledge it is no longer the case.

The heavy elements mercury, thallium, lead, and bismuth are of particular interest to those working in geochemistry and cosmochemistry, because they are the highest atomic number nonradioactive elements in the periodic table. In a 1959 paper (58), we determined the abundances of mercury, thallium, and bismuth in six stone meteorites by the neutron activation technique. As is well-known, the two major advantages of this technique are its high sensitivity and its freedom from contributions to an elemental abundance due to reagent contamination during chemical processing. The latter is of extreme importance for very low abundance elements, especially in the case of mercury owing to the generally high prevalence of mercury contamination in chemistry laboratories.

Previous data on heavy element abundances in stone meteorites at the time this investigation began were very sparse. Suess and Urey stated in their classical paper on elemental abundances that all early analyses of the above elements are suspect because of possible contamination in the laboratory. Three samples of each of six meteorites were examined. The average bismuth content was measured to be 2.2 x 10^{-9} grams of bismuth-209 per gram of stone meteorite. This gives an atomic abundance of bismuth of 0.0016 (based on silicon equal to a million). This value is approximately two orders of magnitude less than the value adopted by Suess and Urey. The average thallium-203 content was measured to be 0.40 x 10^{-9} gram of thallium-203 per gram of stone meteorite. This gives an atomic abundance of thallium-203 of 0.00030. Again, this value is some two orders of magnitude less than the value of 0.0319 adopted by Suess and Urey. The average mercury-202 content was measured to be 30 x 10^{-9} gram of mercury-202 per gram of stone meteorite, giving an atomic abundance of mercury-202 of 0.023, compared with the Suess-Urey value of 0.0846. The results for mercury were based on only five determinations and must be regarded as preliminary.

The values of the cosmic abundances of mercury, thallium, and bismuth are plotted in figure 6. Also plotted are the atomic abundances of thorium-232 (54) and uranium-238 (75). In addition, Suess's and Urey's reported abundances of rhenium, osmium, iridium, platinum, and gold are plotted. This peak in atomic abundances is due to nucleosynthesis via the "r process" where the effect of the 126-neutron shell is experienced at a lower value of the atomic number than that at the beta-stability line. The atomic abundances of thorium and uranium are large compared with the abundances of thallium and bismuth. The fact that chondrites contain at present only a small amount of bismuth relative to thorium and uranium strongly suggests that bismuth has been fractionated from the heavier elements in the past. The longest-lived heavy mass member of the 4n+1 radioactive series is neptunium-237, which has a half-life of only 2.2 million years. Bismuth-209 is the end product of this series. Fractionation of elements with extremely low concentrations in chondrites, such as mercury, thallium, and bismuth, may have occurred prior to or during the formation of the solar system.

Recently conflicting elemental abundances from solar and meteoritic sources are those of iron, chromium, scandium, and europium, the iron/chromium and scandium/europium ratios in particular. Iron falls at a prominent maximum in the abundance curve. Scandium and europium are of particular interest since the odd-even mono-isotope of scandium falls at a major minimum and the two odd-even isotopes of europium comprise a minor minimum in the abundance versus atomic number curve of the elements. With macroscopic amounts of iron present in chondrites, there has seemed little reason to question the conventional wet chemical analysis for iron in chondrites. Hence, if the meteoritic iron/chromium ratio is in error, greater suspicion must be attached to the chromium analyses. Scandium concentrations are reported to approach the 10 microgram per gram

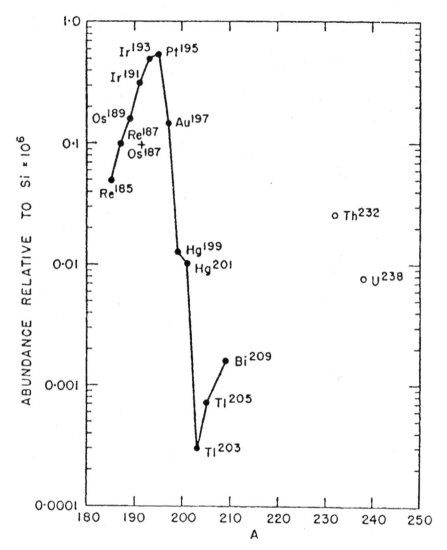

Figure 6. Cosmic abundances of heavy nuclei (58) based on analysis of chondrites. (•) odd A nuclei; (○) even A nuclei. Re, Os, Ir, Pt, and Au from Suess and Urey; Hg, Tl, and Bi (58); Th (54); U (Reed et al., 75). The abundance peak is due to the "r process."

level in chondrites, at which level emission spectrographic data should be reasonably valid. Earlier meteorite europium analyses seemed most susceptible to error since the indicated concentration levels were below one microgram per gram and previous analyses were done many decades ago.

Since the neutron-induced activities of scandium, chromium, and europium all have rich gamma-ray intensities, permitting gamma-pulse-height analysis in favor of greater ease of discrimination against extraneous activities from impurities, and appropriate half-lives, the present work was undertaken to establish the scandium, chromium, and europium abundances in stone meteorites by the independent technique of neutron activation analysis (59). Following the assumptions of Urey discussed previously, our analyses lead to cosmic abundances of 32, 6400, and 0.078 (per million silicon atoms) for scandium, chromium, and europium, respectively. These results are averages for five chondrites studied by simultaneous neutron activation.

The neutron-activation technique was employed also to measure the abundance of ruthenium and osmium (75) in chondrites, which, as reviewed by Suess and Urey, have been deduced hitherto by extrapolation from their measured concentrations in iron meteorites in the absence of direct measurements on chondrites. The mass number distributions of the isotopes of ruthenium and osmium are similar. In each element, a neutron-induced activity is obtainable from a light isotope and from a heavy isotope thereby providing a means of measuring, by neutron activation, the ratios of two isotopes in these elements from various sources.

Some ten chondrites were analyzed and since all are known to be falls, the possibility of contamination by terrestrial weathering is minimized. Other materials examined were two achondrites, a tektite, a trolite, and several terrestrial rocks including the rock standards G-1 and W-1. A lava specimen was obtained from the 1960 Hawaiian eruptions while still hot. These samples serve to give an indication of the scope and range of variations in the abundance of ruthenium and osmium in nature. The average atomic abundances of ruthenium and osmium were measured to be 1.3 and 0.73 (per million silicon atoms). The isotopic ratio of osmium-184/osmium-190 for osmium in meteorites agrees with that in terrestrial material to about one percent, within the limits of experimental error. The experimental uncertainty in determining the ratio of ruthenium-96/ruthenium-102 is greater, and as a consequence the agreement of the ruthenium-96/ruthenium-102 ratio for ruthenium in meteoritic and terrestrial sources cannot be established to much better than two to three percent. With these limits of experimental accuracy, no conclusive evidence for an anomalous isotopic composition of ruthenium and osmium has been found in the samples analyzed in this work.

A 300-gram slice of the Bogou iron meteorite, which fell on August 14, 1962, was quantitatively analyzed 239 days after fall for radioactive nuclei by a nondestructive analysis of the emitted gamma rays (79). Only manganese-54 was positively identified and measured. Upper limits were placed on the presence of several other activities in the meteorite. Unfortunately, the slice of meteorite was obtained too long after fall for the analysis of short-lived activities, which are known to be present in freshly fallen iron meteorites.

B. Fission Research

1. Competition between Fission and Neutron Emission

During the time period of the late 1950s and early 1960s, we expended a considerable effort in the study of the competition between fission and neutron emission as a function of nuclear type and excitation energy. Once a heavy-element compound nucleus is formed, the outcome of the competition between the possibilities of fission, neutron emission, proton emission, and radiation is determined by the relative magnitude of the corresponding widths, Γ_f, Γ_n, Γ_p, and Γ_γ. In our studies, we were interested in the competition in compound nuclei with excitation energy greater than approximately eight million electron volts (MeV). For neutron reactions, this limits the discussion of competition to compound nuclei formed with neutrons of energy at least two MeV and thus eliminates discussion of compound nuclei formed with resonance neutrons. Furthermore, at excitation energies under discussion, the magnitudes of Γ_p and Γ_γ are very small for heavy-element nuclides compared with Γ_f and Γ_n and can be neglected.

The fission process is a collective phenomenon and requires sufficient energy in the form of potential energy of deformation in order to lead to the division of a nucleus. The evaporation of a neutron is, on the other hand, a single-particle phenomenon, which requires an amount of energy at least equivalent to the neutron-binding energy concentrated on one particle at the nuclear surface. Theoretically, therefore, the competition between fission and neutron emission is of great interest. Bohr and Wheeler, in their classical 1939 paper on nuclear fission, show a schematic diagram of the partial transition probabilities for fission and neutron emission from a typical heavy nucleus. For the compound nuclei uranium-239 and thorium-233, respectively, Bohr and Wheeler calculated values of Γ_n / Γ_f in good agreement with present experimental values.

The purpose of our work was to (a) derive all possible Γ_n / Γ_f values from published experimental cross-section data, (b) correlate these Γ_n / Γ_f values with various nuclear parameters, and (c) examine the dependence of Γ_n / Γ_f on excitation energy.

There are four principle types of data available from which one can deduce neutron emission to fission width ratios at excitation energies extending up to approximately 40 MeV. These are fast neutron fission cross sections (45, 48, 51, 74), photofission and photoneutron emission cross sections (13, 24, 31, 40, 45, 48, 51, 74), charged-particle-induced spallation cross sections (51, 74), and charged-particle-induced fission cross sections (68) (limited to nuclei with small fission cross sections).

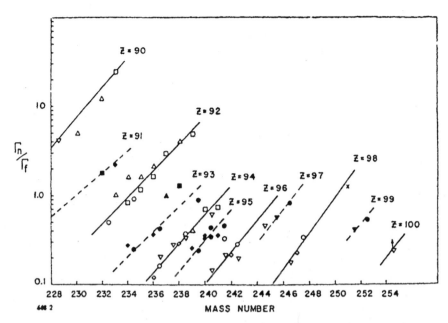

Figure 7. Neutron emission to fission width ratios are plotted as a function of mass number (51). Open and closed symbols refer to fissioning nuclei with even and odd atomic numbers, respectively. Triangles refer to data obtained from photoneutron and photofission experiments and correspond to an excitation energy of 8–12 MeV. Squares refer to data derived from 3-MeV neutron fission cross sections and correspond to an excitation energy of 8–10 MeV. Circles, diamonds, and inverted triangles refer to mean values of Γ_n / Γ_f obtained from spallation excitation functions and correspond to average excitation energies of approximately 13, 18, and 23 MeV, respectively.

Photofission and photoneutron cross sections have been measured for thorium-232 and uranium-238 and the resulting values of Γ_n / Γ_f are independent of excitation energy in the range of 8 to 12 MeV (13, 40) within experimental error. By measuring relative photofission yields (24) obtained with 12-MeV bremsstrahlung and normalizing the data to the $\Gamma_f / (\Gamma_n + \Gamma_f)$ value of uranium-238, neutron to fission width ratios have been derived for a number of nuclides (45). The values of Γ_n / Γ_f determined from photofission and photoneutron emission are plotted in figure 7 as triangles.

Fast neutron fission excitation functions have plateau regions for neutrons of energies between about 2 and 5 MeV. At higher neutron energies the fission cross section of a particular nucleus is observed to rise onto a second plateau. The magnitude of the fission cross section on the first plateau depends directly on the competition between fission and inelastic neutron scattering. For neutron energies above the fission threshold of the target nucleus, the $(n, n'f)$ reaction gives an additional chance for fission to occur and explains the larger fission cross section on the second plateau.

Neutron emission to fission width ratios deduced from 3-MeV neutron fission cross sections represent the competition between neutron emission and fission in compound nuclei (with the mass number increased by one unit over that of the target nuclei) excited to excitation energies of 8 to 10 MeV. Neglecting proton emission and gamma-ray de-excitation, the neutron fission cross section is expressed by $\sigma_f = \sigma_c \Gamma_f / (\Gamma_n + \Gamma_f)$, where σ_c is the cross section for the compound nucleus formation. When σ_f and σ_c are known, $\Gamma_n / \Gamma_f = (\sigma_c / \sigma_f) - 1$. Values of the 3-MeV neutron fission cross sections have been tabulated (45, 51). A theoretical value of 3.3 barns, for the compound nucleus formation cross section of heavy nuclei, was used in calculating Γ_n / Γ_f values. The values of Γ_n / Γ_f from the neutron data are plotted in figure 7 as squares.

When heavy-element nuclides are bombarded with charged particles, the resulting excitation energy is usually high enough so that fission has a chance to compete with neutron emission along an evaporation chain composed of the parent compound nucleus and the various successive daughter nuclei formed by neutron evaporation. For heavy nuclides, this gives rise to a fission cross section, which is a large fraction of the compound nucleus formation cross section. The charged-particle fission cross sections of trans-thorium nuclides are therefore not a very sensitive measure of the competition between neutron emission and fission. A better measure of the competition between neutron emission and fission is obtained by comparing spallation yields of fissionable isotopes with corresponding hypothetical yields had fission not occurred. To deduce the extent to which fission competition has reduced the yield of a particular evaporation product, one needs to know what yield would have occurred without fission competition. For this purpose, we adopted the statistical model of Jackson. In order to correct the Jackson model for fission (51), a fission competition at each step of the evaporation is added in the form of a mean branching ratio $[\Gamma_n / (\Gamma_n + \Gamma_f)]$. The mean values of Γ_n / Γ_f from spallation excitation functions are plotted in figure 7 as circles, diamonds, and inverted triangles, and correspond to average excitation energies of approximately 13, 18, and 23 MeV, respectively.

In figure 7, the neutron emission to fission width ratios are plotted as a function of mass number for the above three types of data. The data fall into rather distinct groups, which define almost straight parallel lines for different values of the atomic number of the compound nucleus. There is no obvious systematic deviation between the Γ_n / Γ_f values derived either from the different types of experiments or from compound nuclei with different excitation energies. A small break in the variation of Γ_n / Γ_f with atomic number occurs at Z of approximately 93. For higher atomic numbers, Γ_n / Γ_f does not depend as strongly on the atomic number as it does for lower atomic numbers.

The strong variation of Γ_n/Γ_f with mass number for a given atomic number can be explained by the following two experimental observations: (1) neutron binding energies increase as the mass number decreases, making it increasingly more difficult to evaporate a neutron, and (2) fission thresholds decrease with decreasing mass number as the fissionability parameter Z^2/A increases (neglecting the reversal in E_f for the very heavy isotopes of a particular element). It might be expected that on a plot of this type, even-even and even-odd compound nuclei would define separate lines, particularly for the cases where only one competition step occurs. Examination of the data from photofission and 3-MeV neutron-induced fission of uranium isotopes, however, shows no systematic deviations from a single line even though both nuclear types are represented. An even-even compound nucleus has a larger neutron binding energy than an even-odd nucleus, but the even-odd product nucleus (following neutron evaporation from an even-even nucleus) has a larger level density than the product nucleus (following neutron evaporation from an even-odd nucleus). The effects of these two factors on Γ_n/Γ_f cancel to a first approximation.

Theoretical expressions for the fission width, Γ_f, and the neutron emission width, Γ_n, as a function of the excitation energy of the compound nucleus have been derived by Bohr and Wheeler and Weisskopf, respectively. By substituting the Fermi gas level density into these expressions and integrating, we obtained equations for numerically evaluating Γ_n and Γ_f (74). In considering the competition between neutron emission and fission, we simplified the latter two equations slightly without loss of accuracy (74), giving the following theoretical ratio of Γ_n/Γ_f,

$$\frac{\Gamma_n}{\Gamma_f} = \frac{4A^{2/3}a_f(E-B_n)}{K_o a_n [2a_f^{1/2}(E-E_f)^{1/2}-1]} \exp\{2a_n^{1/2}(E-B_n)^{1/2}-2a_f^{1/2}(E-E_f)^{1/2}\} \qquad [1]$$

where $K_o = \hbar^2/(gmr_o^2)$, r_o is the nuclear radius parameter, g is a statistical weight factor and equal to 2 for neutrons, and m is the mass of the neutron. The subscripts of a characterize the level density parameter for the fission saddle point, residual nucleus after neutron emission, and the excited compound nucleus. One can correct for even-odd shell effects and pairing energies in an approximate way by replacing B_n by B'_n (where $B'_n = B_n + \Delta$) and E_f by E'_f.

Equation [1] can be greatly simplified by substituting a constant temperature level density for the Fermi gas level density and assuming $T_n = T_f = T$. In this case, the ratio of the neutron emission width to the fission width is given by

$$[\Gamma_n/\Gamma_f = constTA^{2/3} \exp[(E'_f - B'_n)/T] \qquad [2]$$

In this equation we use effective values of the fission thresholds and neutron-binding energies to calculate the energy differences of these two quantities. These effective values differ from the true values because of the dependence of the level density on nuclear type. It is assumed that the exponential level density dependence on excitation energy is determined from a reference mass surface, which differs from the actual ground-state mass because of pairing energies, shells, and so forth.

The dependence of Γ_n/Γ_f on nuclear type is dramatically demonstrated in figure 8A and B where all of the plotted values of Γ_n/Γ_f are derived from data in which only a single competition occurred between neutron emission and fission (51). For figure 8A the abscissa represents the actual difference between the fission threshold and neutron-binding energy $(E_f - B_n)$. In figure 8B the energy difference between the effective values of the fission threshold and neutron-binding energy $(E'_f - B'_n)$ is plotted on the abscissa. The excellent agreement between values of Γ_n/Γ_f for different-type nuclei plotted in figure 8B supports the assumptions made above concerning ground-state masses and fission saddle-point surfaces. The difference between the fission threshold and the neutron-binding energy is a more fundamental parameter for correlating Γ_n/Γ_f values. One sees from figure 8B, a satisfactory correlation of the data with values of Γ_n/Γ_f increasing exponentially, as predicted by Equation [2]. The temperature defines the slope of the data. The excitation-energy dependence on the neutron emission to fission width ratios is difficult to elucidate for trans-thorium nuclides from presently available data. Some comparisons (51) of Γ_n/Γ_f values for the same average fissioning nucleus at different average excitation energies have been made. The only conclusion that can be drawn from these crude comparisons is that Γ_n/Γ_f is not strongly dependent on excitation energy for excitation energies between 8 and 40 MeV. The extraction of the excitation energy dependence of Γ_n/Γ_f from the above comparisons is, however, complicated by the effect on Γ_n/Γ_f of large angular momenta introduced with energetic charged particles (74).

A large angular momentum will reduce the neutron width since part of the excitation energy will be in the form of rotational energy, which results in a reduction of the density of states of the residual nucleus. An approximate correction for this rotational energy, E_R, in the residual nucleus can be made by replacing $(E - B'_n)$ in the Equation [1] by $(E - B'_n - E_R)$. In a similar fashion, one can estimate the effect of angular momentum on the fission width. The effective rotational energy at the saddle, $E_R(f)$, is more difficult to evaluate (74). However, the magnitude of $E_R(f)$ is less than E_R for large values of the angular momentum, and the overall effect of angular momentum is to decrease Γ_n/Γ_f (Γ_f is decreased less than Γ_n).

The nuclei with atomic numbers less than 90 have $[E_f - B'_n]$ values that are much larger than those of the trans-thorium nuclei discussed earlier. As a result the excitation energy dependence of Γ_n/Γ_f (or Γ_f/Γ_n) for nuclides with $Z < 90$ is expected to be very different from those with $Z > 90$. Experimental determinations of Γ_n/Γ_f are not as numerous in this region of the periodic table although many nuclides in the vicinity of lead have been studied with helium-ion projectiles (68). Values of Γ_f/Γ_n for compound nuclei with $Z < 90$ from four helium-ion-induced reactions are shown in figure 9. Values of Γ_f/Γ_n for elements in

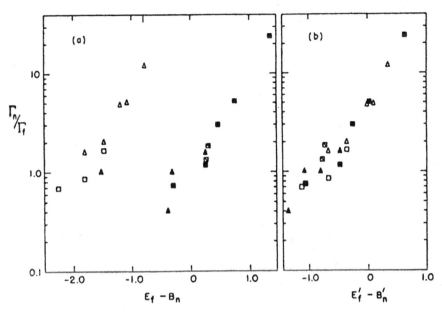

Figure 8. Neutron emission to fission width ratios derived from data in which only a single competition occurred and are plotted as a function of the difference between (A) the actual values of the fission threshold and neutron binding energy $(E_f - B_n)$, and (B) the effective values of the fission threshold and neutron binding energy $(E'_f - B'_n)$. Open symbols refer to even-even compound nuclei, closed symbols refer to even-odd compound nuclei, crossed triangles to odd-even compound nuclei, and crossed squares to odd-odd compound nuclei (51).

this region decrease sharply as the atomic number of the compound nucleus decreases. Values of Γ_f/Γ_n increase sharply with excitation energy, however, as expected from Equation [1] when E_f is much greater than B'_n.

In the late 1950s our group had a scattering chamber constructed and installed on one of the beam lines of the Argonne cyclotron. During this period we were testing different charged-particle detectors suitable for identification and energy measurements of fission fragments. Fortunately, during this time period solid-state detectors became available. They were made by evaporation of metallic gold onto silicon wafers doped with boron and phosphorus. The sensitive area of our initial detectors was approximately six millimeters in diameter. The important advantage of solid-state detectors is that these detectors can be constructed with an almost negligible window thickness, while the depth of the sensitive region of the detector can be adjusted by the magnitude of the reverse bias potential to correspond to the range of the fission fragments. Thus particles with longer ranges, such as scattered alpha particles, deposit only a small fraction of their energy in the sensitive region of the detector. This leads to a larger

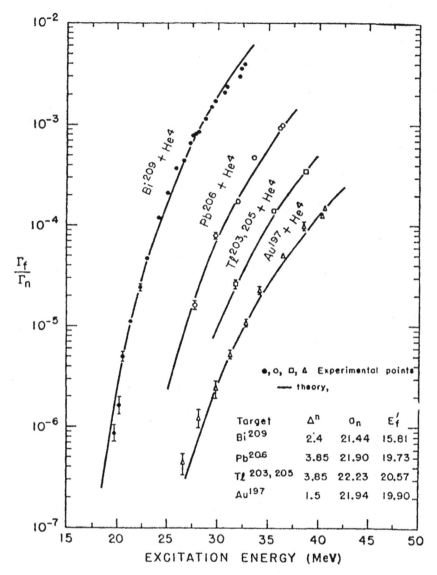

Figure 9. Fission to neutron emission width ratios for compound nuclei with Z < 90 plotted as a function of excitation energy. The level density parameter of the fissioning transition nucleus, a_f, is assumed to be A/8. The value of $B'_n = B_n + \Delta^n$. The values of a_n and E_f which give a good fit to the data, are listed (68).

separation of the pulse heights for the fission fragments from the undesirable background from other particles. Such detectors were used to measure the fission cross sections described in figure 9.

The novelty of these solid-state detectors in the late 1950s is illustrated by the following series of events. On December 29, 1959, I was invited to give a lecture entitled "Recent Experimental Studies of Nuclear Fission" at the annual meeting of the American Association for the Advancement of Science (AAAS), which was held in a hotel in downtown Chicago. During the lecture I circulated one of our solid-state detectors through the audience and to my dismay someone stole it! Although people in our laboratory were manufacturing solid-state detectors for us, this particular detector was given to us for testing by an industrial firm, a common procedure at that time. Hence, I was particularly embarrassed to have it stolen. Questions of patents and lawsuits crossed my mind. The next morning Earl Ubell, science editor of the New York *Herald Tribune,* published a feature article about my lecture entitled, "Small, Cheap Gadget Counts Atoms Better." Ubell speculated that the tiny new wafers, in addition to their use in physics, would revolutionize the measurement of radioactivity in the field of biology. He went on to state that "now biologists will be able to sew the tiny wafers into almost any organ and measure the radioactivity there in the living animal." My lecture at the AAAS meeting in Chicago and the subsequent newspaper publicity brought me a number of letters from university and industrial scientists interested in the use of silicon detectors for the characterization of gamma rays and charged particles.

On January 16, 1958, I received a letter from Mr. P. W. McDaniel, Acting Director of the Division of Research for the Atomic Energy Agency (AEC) to testify before Congress on the AEC's physical research program during the first two weeks of February, 1958. I was asked, in particular, to discuss the research I was conducting on the "studies of the fission process." This was an exhilarating, although scary, experience. It was particularly challenging to formulate answers to Congress members' questions at the level that was both appropriate and satisfactory.

On February 4, 1958, Professor Niels Bohr, along with his fourth son, Aage, visited Argonne. Niels Bohr was 73 at this time and Aage was 35 years old. Bohr had six sons and Aage was the only son following his father into theoretical physics. Louis Turner, then Director of the Argonne Physics Division, invited me along with a small group of people to participate in an informal discussion with Bohr around a table in a small conference room. It was fun and a great honor to have the opportunity to discuss nuclear fission with Bohr, who had written the classical theoretical paper on the subject with John Wheeler in 1939. In 1922, Niels Bohr won the Nobel Prize in Physics, and he made Copenhagen one of the leading centers of research in nuclear physics. By the time of my sabbatical at the Bohr Institute in 1974, Niels Bohr was no longer living (he died in 1962), however, we met his wife, Margrethe (to be described later).

Another significant event occurred in March, 1958, when I received a letter from the laboratory director informing me that I had been promoted to senior scientist.

The Gordon Conference on Nuclear Chemistry had its initial meeting in June, 1952, with Glenn T. Seaborg as chairman. At the 1957 meeting, I was elected to be chairman of the 1958 Conference on Nuclear Chemistry, which meant that I had the responsibility for organizing the scientific program and selecting the speakers. Five of the nine half-day sessions were devoted to subjects of special interest to me: Photonuclear Reactions, Fission and Neutron Widths, Nuclear Reactions and Geochemistry, Cosmic Abundance of the Elements, and Synthesis of the Elements. I was able to attract a distinguished group of participants, including five Nobel Prize winners. Two of the members of my research group were speakers, one in the session on Fission and Neutron Widths and one in the session on Cosmic Abundances. I wrote to the president of the Soviet Academy of Sciences informing him that I had invited three Soviet scientists to attend our conference. The small New Hampshire villages, where the Gordon Conferences are held in local college facilities, had some problems cashing checks from Soviet banks! I attended every Nuclear Chemistry Gordon Conference from the initial one in 1952 until my retirement, except for the three years I was on sabbatical leave in Europe.

In the summer of 1958, the Department of State designated me to be a technical advisor to the US Delegation to the Second International Conference on the Peaceful Uses of Atomic Energy to be convened in Geneva, Switzerland, on September 1, 1958. During the meeting I presented a paper on the "Nuclear Fission Process." This was an excellent meeting in a beautiful city, which by now I had visited several times. Leading scientists from many nations were in attendance, and there were numerous occasions to have interactions with them. After the conference, I visited laboratories in Belgium, The Netherlands, and England and delivered invited lectures.

2. Fission Fragment Angular Distributions

During this time period, we began to do a variety of on-line studies of nuclear fission as solid-state detectors became available. These detectors were especially advantageous in the study of fission fragment angular distributions. We investigated such angular distributions for fission induced with a variety of projectiles and at a range of saddle-point excitation energies extending from zero to tens of MeV.

In 1956, Aage Bohr suggested that low-energy fission may be understood in terms of a very few levels in the transition-state nucleus (or saddle point). Although the level spacing in the compound nucleus as an excitation energy of about 6 MeV is of the order of one electron volt or less, most the of excitation energy goes into deformation energy during the passage from the initially excited compound nucleus to the highly deformed transitional-state nucleus. Hence, the transition-state nucleus is thermodynamically "cold" and is expected to have a spectrum of excited states analogous to that of a normal nucleus near its ground state, except for effects due to its much larger deformation.

Information about the nuclear states in the highly deformed transition-state nucleus can be derived from fission fragment angular distributions. From the excitation energy dependence of the angular distributions, it is possible to characterize the saddle states in terms of their energies and quantum numbers J (spin) and K (projection of J on the nuclear symmetry axis of the transition-state nucleus). If we assume that the fission fragments separate along the nuclear symmetry axis and that K is a good quantum number in the passage of a nucleus from its transition state to the configuration of separated fragments, the directional dependence of fission fragments resulting from the transition state with quantum numbers J, K, and M is uniquely determined. The quantum numbers J and M (projection of J on the space-fixed axis which is usually taken as the beam direction) are conserved in the entire fission process. Whereas J and M are fixed throughout the various extended shapes on the path to fission, no such restriction holds for the parameter K. The K value (or values) of the transition nucleus are unrelated to the initial K values of the compound nucleus. Once the nucleus reaches the transition-state deformation, K is a good quantum number beyond this point of the fission process. The relationship between J, M, and K is schematically illustrated in figure 10. For actinide nuclei where the first or inner barrier is the higher energy barrier, we assume that K mixing occurs prior to reaching the second or outer barrier. At the deformation of the inner barrier, heavy nuclei are very soft toward axially asymmetric deformations which mix K. On the basis of theoretical expectations and limited experimental evidence, we assume that the K distribution is frozen in at the second barrier for low-energy fission (nuclei in the lead region have a single barrier). At high excitation energies, the K distribution of the liquid-drop saddle is expected to be applicable.

The differential fission fragment cross section for a particular channel (J, π, K, M) at angle θ is given by:

$$\frac{d\sigma_f}{d\Omega}(J,\pi,K,M,\theta) = [\sigma_f(J,\pi,K,M)/2\pi][(2J+1)/2]\,|\,d^J_{M,K}(\theta)\,|^2 \qquad [3]$$

where the $d^J_{M,K}(\theta)$ functions are defined in (87), and the differential fission fragment cross section is to be expressed in the same units per steradian as the total fission cross section. The factor 2π appears instead of the usual 4π due to the fact that two fragments separated by 180° arise from each fission event.

We studied fission fragment angular distributions at very-low excitation energies in the transition-state nucleus by direct reactions induced by charged particles and with low-energy monoenergetic neutrons. I will first describe a study of even-even transition state nuclei excited by the (α, α') reaction. This is an important method for investigating the saddle states in an even-even nucleus

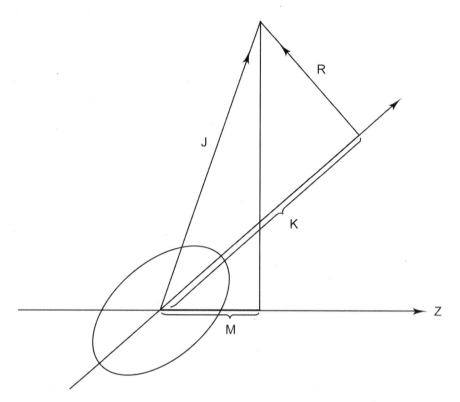

Figure 10. Momentum coupling scheme for a deformed nucleus. The vector J defines the total angular momentum. The quantity M is the component of the total angular momentum on the space-fixed Z axis. We define this direction as the beam direction. The quantity K is the component of the total angular momentum along the nuclear symmetry axis. The collective rotational angular momentum R is perpendicular to the nuclear symmetry axis; thus, K is entirely a property of the intrinsic motion. The angle θ in this chapter represents the angle between the nuclear symmetry axis and the space-fixed Z axis (138).

since for this case both the projectile and the target have spin zero. A depleted uranium-238 target was bombarded with an external cyclotron beam of 43-MeV helium-4 ions (83). The angular correlation between the coincident scattered alpha particles and the fission fragments was measured for two fixed angles of the α' detector, 45 degrees and 60 degrees to the beam direction. An Argonne multiparameter analyzer was used to record three bits of information for each coincident event: (1) the scattered alpha particle (α') energy; (2) the fission-fragment energy; and (3) the time delay between the coincident α' particle and the fission fragment. From the measured α' energy and the reaction kinematics, the nuclear excitation, E, can be calculated. By recording the time spectrum, pulses from the different r_f cycles can be discerned, enabling one to unfold the true events from the chance events. With the α' detector fixed at either 45 or 60 degrees, the maximum coincidence rate for an energy bin corresponding to 5.6 to 6.2 MeV of excitation energy in the compound nucleus coincided with the inelastic recoil axis, θ_R. This result confirms that the angular momentum symmetry axis is identical to θ_R for the above α' angles.

The lowest energy state of a deformed even-even nucleus at the saddle point is expected to have quantum numbers J=0+, K=0 analogous to the less deformed ground state. Built upon this state one would expect to observe collective rotational states with energy:

$$E_R(J) = (\hbar^2/2 \mathcal{I}_{\perp \text{saddle}}) J(J+1) \qquad [4]$$

where J=0, 2, 4, 6 . . . , and $\hbar^2/2 \mathcal{I}_{\perp \text{saddle}}$ is calculated for a rigid body from liquid-drop theory to be approximately 2 keV for the transition state of uranium-238. With the α' energy resolution, rotational members of the K=0 band are irresolvable, however, it is possible to resolve different K bands.

The relative differential fission cross section for the above case (where K=M=0 and positive parity) is given by

$$\frac{d\sigma_f}{d\Omega}(\theta) \propto \sum_J (2J+1) b_J \, | \, P_J(\cos\theta) \, |^2 \qquad [5]$$

where b_J is the relative probability of fissioning through a specific J state and $|P_J(\cos\theta)|$ are Legendre polynomials.

The angular distribution of fission fragments passing through channels within 600 keV of the fission threshold is shown in figure 11. The experimental points have been corrected for the finite solid angle of the detectors. The very large experimental zero degrees to 90 degrees anisotropy of 7 is characteristic of fission through levels where K=0 and $M << J$. Because the exact form of b_J for this (α, $\alpha' f$) reaction is not known, the solid curve shown in figure 11 is an empirical fit to the data using Equation [5] with b_J=exp-$(J/7)^2$ and summing only even values of J with M=0. As can be seen from figure 11, an excellent theoretical fit to the fission fragment angular distribution is obtained with a single K=0+ band in the transition nucleus. It is possible to make a definite determination of whether a band is K=0+ or 0- by a careful investigation of the (α, $\alpha' f$) angular correlation near 90 degrees (see the small peak in figure 11 at 90°). The yields for the various even-J positive parity states reinforce each other at 90 degrees, whereas the odd-J negative parity states all have zero yield at this angle.

The fission fragment angular distribution in figure 11 has particular significance. The very large experimental anisotropy conclusively shows that the states in the transition nucleus control the fission fragment angular distribution. In addition, K is a good quantum number as the fission process moves from the saddle deformation to fission fragments. The maximum excitation energy of

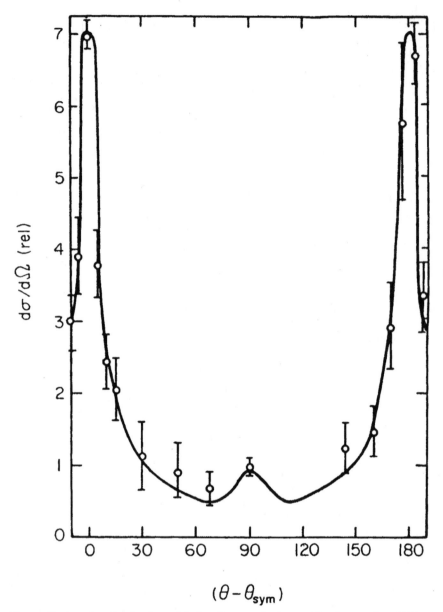

Figure 11. Angular distribution of fission fragments observed in the uranium-238 $(\alpha,\alpha'f)$ experiment for excitation energies of 5.6 to 6.2 MeV (zero to 0.60 MeV of excitation energy in the transition nucleus). The angles are measured relative to the angular momentum symmetry axis, θ_{sym}, which in this case was measured to be within experimental error identical to the recoil axis (83).

the transition nucleus of 600 keV in figure 11 is insufficient to break nucleon pairs in the transition nucleus (although the energy deposited in the target nucleus was up to 6.2 MeV, some 5.6 MeV of this energy went into deformation energy at the transition-nucleus deformation).

Excitation of low-lying states in transition nuclei excited by neutron capture and the (d, p f) direct reaction will be discussed later.

In the previous paragraphs, a theory of fission fragment angular distributions was described for isolated transition states at low excitation energies with specific quantum numbers. Next I will extend this theory to higher excitation energies where the transition levels can be described by statistical methods. The density of levels in the transition nucleus with spin J and projection of J on the nuclear symmetry axis equal to K is given by the approximate constant temperature level density:

$$\rho(J,K) \propto \exp\ [(E - E_{rot}^{J,K})/t] \qquad [6]$$

where E is the total energy, $E_{rot}^{J,K}$ is the energy tied up in rotation for transition state (J, K) and t is the thermodynamic temperature. The derivation of Equation [6] requires the constancy of t for small changes in excitation energy around E. The thermodynamic energy available to the nucleus is the quantity $(E - E_{rot}^{J,K})$. The rotational energy of a nucleus in its saddle-point deformation is (see figure 10),

$$E_{rot}^{J,K} = (\hbar^2 / 2\ \mathscr{I}_\perp)(J^2 - K^2) + (\hbar^2 / 2\ \mathscr{I}_\|)K^2 \qquad [7]$$

where $\mathscr{I}_\perp$ and $\mathscr{I}_\|$ are nuclear moments of inertia about axes perpendicular and parallel to the symmetry axis, respectively. Substitution of Equation [7] into Equation [6] gives,

$$\rho(J,K) \propto \exp\{(E/t) - (\hbar^2 J^2 / 2\ \mathscr{I}_\perp t) - (\hbar^2 K^2 / 2t)\ [(1/\mathscr{I}_\|) - (1/\mathscr{I}_\perp)]\} \qquad [8]$$

For fixed values of E and t the number of transition levels $\rho(J,K)$ depends on two quantities, $(\hbar^2 J^2 / 2\mathscr{I}_\perp t)$ and $(\hbar^2 K^2 / 2t)\ [(1/\mathscr{I}_\|) - (1/\mathscr{I}_\perp)]$. If in addition J is fixed, then the distribution of K becomes,

$$\rho(K) \propto \exp\{-(\hbar^2 K^2 / 2t)\ [(1/\mathscr{I}_\|) - (1/\mathscr{I}_\perp)]\} \qquad [9]$$

Equation [9] is equivalent to a Gaussian K distribution,

$$\rho(K) \propto \exp(-K^2/2K_o^2) \qquad K \leq J$$
$$= 0, \qquad K > J \qquad [10]$$

where $K_0^2 = t/\hbar^2[(1/\mathscr{I}_{\parallel})-(1/\mathscr{I}_{\perp})]$. If the quantity $[(1/\mathscr{I}_{\parallel})-(1/\mathscr{I}_{\perp})]$ is replaced by $1/\mathscr{I}_{\text{eff}}$, then

$$K_o^2 = t\mathscr{I}_{\text{eff}}/\hbar^2 \ . \tag{11}$$

For a Gaussian K distribution, an exact expression for the fission fragment angular distribution may be derived from Equation [3] by proper weightings of J, M, and K. The exact and several approximate theoretical expressions are given and their merits are discussed in detail elsewhere (108, 138). In the limit when the target and projectile spins are zero and no particle emission from the initial compound nucleus occurs before fission (i.e., $M=0$), the angular distribution, when many J values of the compound nucleus contribute and the transmission coefficients T_l are known, becomes

$$W\theta \propto \sum_{J=0}^{\infty}(2J+1)T_J \sum_{K=-J}^{J} \frac{(2J+1)|d_{M=0,K}^{J}(\theta)|^2 \exp(-K^2/2K_o^2)}{\sum_{K=-J}^{J}\exp(-K^2/2K_o^2)} \tag{12}$$

where the transmission coefficients are written as T_J since $l=J$ when $M=0$. Examples of reactions, with target and projectile spin both zero, are alpha-particle-induced fission of even-even target nuclei (limited to reactions where only first-chance fission occurs, since M is no longer zero if fission follows initial de-excitation by neutron emission).

Fission fragment angular distributions were studied for a number of targets bombarded with helium ions (65, 69, 86), deuterons (76) and protons (77). In addition fission fragment anisotropies, such as W (174°) / W (90°), were studied as a function of projectile energy in small energy steps. Some of the more striking features of the experimental observations are summarized in the following statements:

a) The fission fragments have the largest differential cross sections in the directions forward and backward along the beam in compound nucleus reactions (in direct reactions along the symmetry axis).

b) The anisotropies are largest for the heaviest projectile and smallest for the neutron and proton reactions.

c) The anisotropy increases whenever a threshold is reached at which it becomes energetically possible for fission to occur in the residual nucleus that is left behind after the evaporation of some definite number of neutrons (e.g., alpha-particle-induced fission of radium-226 shows this effect).

d) The anisotropies are approximately the same for odd-A and even-even targets.

e) The anisotropy decreases as the values of Z^2/A increases.

The angular distribution of fission fragments depends on two quantities: the angular momentum brought in by the projectile and the fraction of this angular momentum that is converted into orbital angular momentum between the fragments. This fraction is characterized through a parameter K, where K is defined as the projection of the angular momentum J on the nuclear symmetry axis (see figure 10). The distribution of angular momenta of the fissioning system may be controlled by the experimenter through his choice of the projectile and its energy. The distribution of K, however, is intrinsic to the nucleus itself. From measurements of fission fragment angular distributions, it is possible, therefore, to deduce the K spectrum of states at the saddle-point deformation.

The calculation of K_0^2 from the fission fragment angular distribution (or the anisotropy) by a relation such as Equation [12] requires projectile transmission coefficients T_l, which are computed with optical model theory. As an example, the values of K_0^2 derived for helium-ion-induced fission of bismuth-209 are plotted in figure 12 as a function of the excitation energy of the transition-state nucleus in MeV (87). Since bismuth-209 has a large spin of 9/2, the exact theory was used to deduce K_0^2 rather than the approximate Equation [12]. The solid line in figure 12 is a theoretical fit of the K_0^2 data. The behavior of K_0^2 is approximately accounted for by statistical theory with the equation of state $E=a_f t^2 - t$ and a constant effective moment of inertia at all excitation energies ($\mathscr{I}_{\text{eff}}$ in this case is approximately equal to two-thirds of $\mathscr{I}_{\text{sph}}$, which results in a highly deformed transition-state nucleus). The interpretation of the energy dependence of the anisotropy of heavy elements beyond bismuth is complicated due to multichance fission.

Modification in the statistical theory to account for nuclear pairing was investigated (65, 87). The modifications of the independent particle model by the pairing interaction are especially relevant for fission fragment angular distributions, because they depend directly on the effective moment of inertia. This relationship is parameterized through K_0^2. Because of the pairing interaction of single particle states, the particle excitations, which contribute to the total projection on the symmetry axis, do not extend continuously to the ground state. For a given excitation energy a superfluid nucleus has a significantly reduced K_0^2. Hence, the dependence of K_0^2 upon excitation energy is a rather sensitive method of distinguishing between the superconducting model and the Fermi-gas-independent-particle model.

To investigate the dependence of fission-fragment anisotropy on the fission-fragment mass, researchers must choose fissioning systems in which single-chance fission is dominant. This is the case for fissioning nuclei in the vicinity of lead, even at rather high excitation energies. The fissioning systems investigated in the lead region were 42-MeV helium-ion-induced fission of lead-206 and bismuth-209. The anisotropies of fission-fragment isotopes of elements ruthenium, palladium, silver, strontium, molybdenum, and bromine were measured (82).

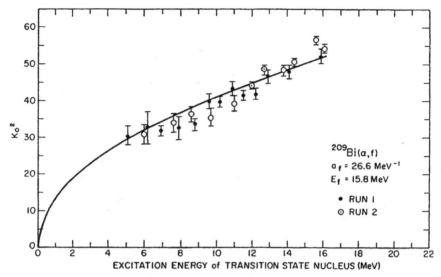

Figure 12. Excitation energy dependence of K_0^2 deduced from helium-ion-induced fission of bismuth-209 for the assumption of single chance fission. The abscissa is the excitation energy above the fission barrier. The solid line represents the theoretical energy dependence predicted by statistical theory (87).

The anisotropy is independent of mass for a mass ratio of 1.0 to 1.3. However, the anisotropy of bromine-83 (M_1/M_2=1.53) from the helium-ion-induced fission of bismuth-209 is some ten percent less than other fragments. This may be associated with an increase in $\mathscr{I}_{\text{eff}}$ for a highly deformed configuration containing a 50-neutron shell.

Measurements of the fission fragment anisotropy as a function of mass asymmetry for heavy elements must be done at low energy, where only single-chance fission occurs. No detectable dependence of the anisotropy on mass asymmetry was observed for heavy element fission (90).

3. Fission Fragment Kinetic Energy Distributions

The kinetic energy of the fission fragments arises from the Coulomb repulsion of the two charged fragments. Therefore, one expects the kinetic energy to be proportional to a Coulomb energy $Z_1 Z_2 e^2/r$, where r is the separation distance of the centers of charge of the two fragments. Setting $Z_1=Z_2$ and using the proportionality $r \propto A^{1/3}$, one expects a correlation of the kinetic energy release with $Z^2/A^{1/3}$.

Single fragment and total kinetic energy distributions were measured for fission induced by helium ions and deuterons on targets of gold, thallium, lead, and bismuth (72). The energy distributions of the fragments were observed using surface barrier solid-state detectors having a sensitive area of about 0.3

cm^2. The energy calibration of the detectors was performed with a californium-252 spontaneous fission source. The average total kinetic energy release for 43-MeV helium-ion-induced fission of the above four targets, were 138 ± 4, 143 ± 5, 146 ± 5, and 148 ± 4 MeV, respectively. For deuteron fission of bismuth, the average total kinetic energy release was 143 ± 5 MeV. The indicated errors include uncertainties arising from calibration of the detectors and target thickness uncertainties.

The results from these studies show that the kinetic energies for symmetric fission in the lead region correlate well with the $Z^2/A^{1/3}$ parameter, as do the kinetic energies of heavy elements that fission asymmetrically. Present evidence indicates that the kinetic energy release in fission is fairly independent of excitation energy. It is of interest to deduce possible scission configurations from the total kinetic energy releases. It is reasonable to assume that all of the kinetic energy arises from the Coulomb interaction energy of the two fragments at the "scission configuration." It is also reasonable to approximate the scission configuration for symmetric fission by two equal tangent prolate spheroids described by major and minor semi-axes C and A. From the kinetic energies of the above four fissioning systems, the ratio of C/A for each of the two prolate spheroids at the scission configuration (72) is calculated in all cases to be 2.17 ± 0.10.

The kinetic energies of coincident fission fragments were measured for helium ion induced fission of bismuth-209, radium-226, and uranium-238 with gold-surface barrier detectors and an Argonne three parameter analyzer (78.) The data obtained at each bombarding energy were analyzed to give total kinetic energy and mass yield distributions for fixed values of the average total kinetic energy, and the total kinetic energy distributions as a function of heavy fragment mass M_H. The mass yield distributions for bismuth-209, uranium-238, and radium-226 are single, double, and triple-humped, respectively, as shown in figure 13. The total kinetic-energy release for bismuth-209 fission decreases smoothly with increasing values of M_H and its variance remains constant. The total kinetic energy as a function of M_H for all the bombardments of radium-226 and uranium-238 show structure with a maximum kinetic energy at M_H=135, while each bombardment gives a maximum in the variance of the total kinetic energy at M_H=131. This structure in the kinetic energy for radium-226 and uranium-238 is interpreted in terms of shell structure in the heavy fragment. The larger kinetic energy observed for symmetric fission of heavy elements with energetic projectiles over that observed for thermal-neutron fission is assumed to be due to a smaller effective separation of charge centers at the scission configuration. This effect may possibly result from a temperature-dependent viscosity and tensile strength of the nuclear fluid and leads to the interesting speculation that information on these parameters may be inferred from nuclear fission data. The full-width at half-maximum height in the total kinetic-energy distribution for 42-MeV helium-ion-induced fission of a bismuth-209 target is 16 ± 1 MeV, in excellent agreement with a theoretical calculation of Swiatecki and Nix.

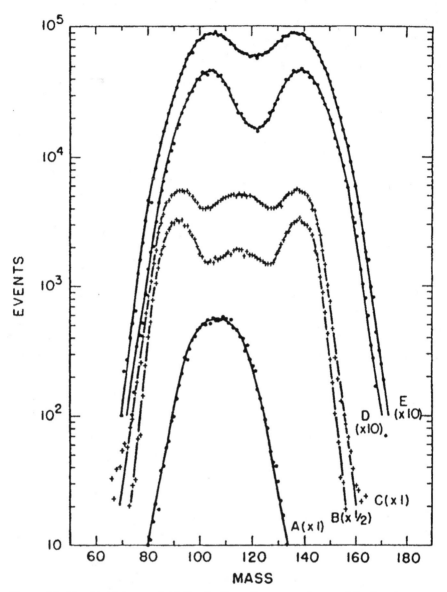

Figure 13. Total initial mass-yield distributions. For convenience of display, the number of events for each system has been multiplied by a scale factor given in parenthesis. (A) 42.0-MeV helium-ion-induced fission of bismuth-209 (B) 30.8-MeV helium-ion-induced fission of radium-226 (C) 38.7-MeV helium-ion-induced fission of radium-226 (D) 29.4-MeV helium-ion-induced fission of uranium-238 (E) 42.0-MeV helium-ion-induced fission of uranium-238 (78).

C. Nuclear Reactions

The cross sections for the (p,γ), (p,n), and (p,2n) reactions were measured for a bismuth-209 target bombarded with protons with energies up to 10.65 MeV (34). The resulting polonium-210, polonium-209, and polonium-208 alpha-particle activities were counted off-line. The threshold for the (p,2n) reaction was determined to be 9.65 MeV. The total experimental reaction cross section for protons on a bismuth-209 target agrees with theory.

Cross sections for the (α,2n), (α,3n), and (α,4n) reactions were measured for a bismuth target bombarded with helium ions with energies from 20.6 to 43.3 MeV (56). In addition, cross sections for the (d,p), (d,n), (d,2n), and (d,3n) reactions were measured for a bismuth target bombarded with deuterons with energies from 6.3 to 21.5 MeV (56).

The resulting radioactive nuclides were analyzed off-line following each bombardment, and the alpha-active nuclei were identified by their characteristic alpha-particle energies with an ionization chamber and a pulse-height analyzer. The experimental results of the deuteron and helium-ion-induced reactions show the general features of a compound nucleus formation mechanism, that is, a rapid rise of cross sections of a given mode of neutron emission with increasing bombarding energy above its threshold, and a subsequent rapid drop of cross section at higher energies due to competition with emission of one more neutron. These (d,xn) and (α,xn) compound-nucleus-like reactions were compared with the predictions of the Jackson model and found to be in good agreement. The (d,n) and (d,p) excitation functions, however, do not have the characteristic compound-nucleus peak. The cross sections for these reactions are essentially constant with deuteron energy from 12 to 21.5 MeV, displaying a direct-reaction mechanism known as the nuclear-stripping process.

Excitation functions of reactions formed by bombardment of uranium-235 and uranium-238 targets with helium ions of energies from 18.0 to 43.0 MeV and deuterons of energies from 5.8 and 21.5 MeV were studied (57). The yields of the products from each spallation reaction were determined off-line by radioactivity or mass spectrometric techniques. The total fission cross sections were obtained from the fission-product barium-140 activity. The excitation functions for these uranium targets are compared with similar excitation functions of nonfissile nuclei to show the effect of fissionability on spallation cross sections. Mean values of Γ_n/Γ_f are derived for several heavy nuclides by fitting the experimental excitation functions with theoretical cross sections calculated from the Jackson model modified for fission. The sum of the spallation cross sections for uranium-238 is considerably larger than that for uranium-235. For the helium-ion-induced reactions on a uranium-238 target, the fraction of the total reaction cross section going into spallation products decreases from 70 to 10 percent as the helium-ion energy increases from 20 to 40 MeV.

The total reaction cross sections is given by:

$$\sigma_R = \pi \lambdabar^2 \sum_0^\infty (2l + 1) T_l(\varepsilon) \qquad [13]$$

where $\lambdabar$ is the de Broglie wavelength, l is the angular momentum of the incident particle in units of $\hbar$, and $T_l(\varepsilon)$ is the transmission coefficient of the incident particle of energy ε. In the late 1950s, I was interested in comparing our experimental helium-ion-induced reaction cross sections with theory. However, at that time alpha-particle transmission coefficients with a complex nuclear potential were not available. George Igo of the Lawrence Berkeley Laboratory had developed a computer program and calculated total reaction cross sections with a complex nuclear potential for a few targets with alpha-particle projectiles of a few energies. Hence, I decided to collaborate with Igo in order to calculate alpha-particle transmission coefficients for a large variety of targets at a number of alpha-particle energies with a complex nuclear potential.

Arrangements were made between the Argonne National Laboratory and the Lawrence Berkeley Laboratory for me to spend the summer of 1960 in Berkeley. We rented our house to summer guests at Argonne and packed necessities and family in our Pontiac station wagon. A couple weeks of vacation time were taken as we camped and enjoyed sightseeing on our way from Chicago to Berkeley. Scientific stops were made in Los Alamos, San Diego, and Irvine as well as pleasure stops in National Parks and Disneyland. The Berkeley National Laboratory found us a comfortable house on Shattuck Avenue in Berkeley. At the Laboratory I shared an office with John Wheeler, who also was spending the summer in Berkeley. This was fortunate for me because of Wheeler's interest in fission (as I mentioned previously, he co-authored with Niels Bohr the classical theoretical paper on Nuclear Fission in 1939).

My summer in Berkeley was very productive. Computing time was available to me on the Laboratory's largest computer. Transmission coefficients, T_l, and total reaction cross sections, σ_R, for alpha particles in the energy range 0 to 46 MeV interacting with 20 target nuclei, with atomic numbers ranging from 10 to 92, were calculated with an optical model program utilizing a complex nuclear potential. The dependence of the values of the transmission coefficients and, hence, the total reaction cross section on the Woods-Saxon parameters was investigated as a function of projectile energy. In addition, the optical model reaction cross sections were compared with those derived from (a) a square-well potential and (b) a model which approximates the real optical model potential barrier by a parabola and makes use of the Hill-Wheeler penetration formula for a parabolic potential. These general results are published in the open literature (67), while the tables of transmission coefficients are published in an Argonne National Laboratory report (ANL-6373). These tables of transmission coefficients were very popular as hundreds of requests were received for copies of this report.

After our summer in Berkeley, we traveled north along the coast, stopping to give colloquia at Oregon State University and the University of Washington in Seattle. While camping at Crater Lake National Park in Oregon, we had a very scary experience. Long after midnight we heard loud screaming. First thinking it was some rowdy teenagers having fun, we ignored the noise. The screams continued, however, and became more desperate sounding. We decided we had to investigate. It was late in the season, with few campers. Hence, there was a scarcity of food in the garbage cans for the bears. In 1960, we had a tent which, coupled to our station wagon, slept six people. We uncoupled the tent, packed all six of us into the station wagon, and drove in the dark in the direction of the screaming. We soon came upon a man with a small child in his arms and blood streaming from his face. We rushed them to an emergency station, some ten miles down the mountain, which we had noticed on entering the park. We learned they had a small trailer containing food parked near their tent. In the night when hearing the bear attack the trailer, the man pointed a flashlight at the bear from his tent, hoping to scare the bear away. Instead, the bear turned and rushed toward the light and mauled his face. The next day on the road, we heard the story on our car radio and learned the man had lost one eye. After staying in a hotel in Seattle, we once again tried camping in Glacier National Park. However, in the night we again heard footsteps of bears outside. That was the end of our camping on our return trip to Chicago! In Yellowstone National Park, we opted to stay in the park cabins and used motels thereafter.

Once I was back at work at Argonne, the total reaction cross sections were measured for targets of uranium-233 and uranium-238 bombarded with 18 to 43 MeV helium ions (66). The total reaction cross sections were obtained by summing the combined excitation functions of all reaction products. The targets uranium-233 and uranium-238 were chosen because of their large fission cross sections. In addition, the spallation cross sections are known (57), and represent a small fraction of the reaction cross section. The fission cross sections of these targets make up most of the reaction cross sections and are measured with good accuracy by a technique that utilizes solid-state detectors for measuring the fission fragments. The total reaction cross sections for a uranium-233 target bombarded with helium ions of energies from 18 to 43 MeV are shown in figure 14. As shown in this figure, the data are in excellent agreement with the theoretical reaction cross sections calculated with an optical model employing a complex nuclear potential (67).

Photofission of heavy element targets was induced with mono-energetic gamma rays that were produced by proton bombardments of fluoride targets (70). Protons of variable energy were accelerated in the Argonne 4-MeV Van de Graaff generator. The gamma rays used in these experiments were obtained by bombarding thin fluoride targets with either 1.56- or 2.10-MeV protons. At the lower bombarding energy, the predominant gamma-ray component is the

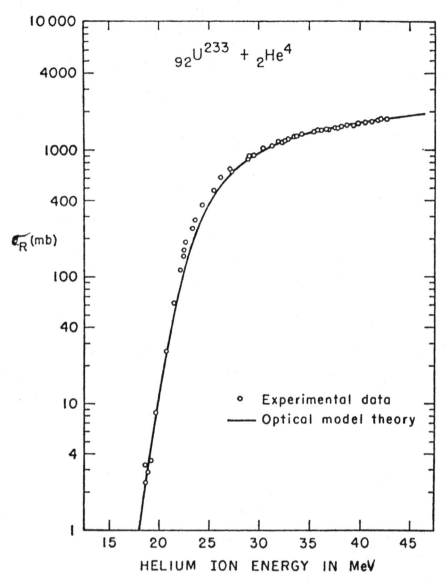

Figure 14. Comparison of experimental and theoretical total reaction cross sections for uranium-233 bombarded with helium-ions. The theoretical cross sections are calculated with an optical model employing a complex nuclear potential (66).

6.14-MeV line. At the higher proton bombarding energy, the 6.91- and 7.12-MeV gamma-ray groups are in greater abundance. The heavy-element targets were contained in a cancellation ionization chamber of 2π geometry. The spectrum and the intensity of the gamma radiation incident upon the fissionable targets were monitored by a high-resolution scintillation spectrometer employing an anticoincidence annulus. Both the fission chamber and the scintillation spectrometer were mounted on the axis of the incident proton beam such that the gamma beam traversed first through the fission chamber and then into the gamma-ray detector. The external beam windows and internal electrodes of the fission chamber were such that any attenuation of the transmitted gamma-ray beam was negligible.

The photofission cross sections of thorium-232, uranium-238, uranium-236, uranium-235, uranium-234, uranium-233, and neptunium-237 for a 7.0-MeV average gamma-ray energy (a mixture of the 6.91- and 7.12-MeV gamma rays) are 9, 15, 28, 33, 52, 44, and 45 millibarns, respectively (70). The 7.0-MeV photon absorption cross section of heavy nuclei was determined to be 60 ± 20 millibarns. From this cross section one deduces that the gamma ray strength function, Γ_0/D, for heavy nuclei at 7.0 MeV is $(2.5 \pm 0.8) \times 10^{-4}$.

Experimental excitation function data for (α,xn) reactions on targets of lead-206, lead-208, bismuth-209, and uranium-238 as well as $(n,2n)$ reactions on targets of thorium-232 and uranium-238 were analyzed by a Monte Carlo technique to investigate the dependence of the nuclear level density on excitation energy (63). The analysis supports the prediction of the completely degenerate Fermi gas model that the nuclear temperature varies as the square root of the excitation energy. The value of the level density parameter, a, deduced from the experimental data is changed significantly when shell and pairing effects are taken into account by displacement of the characteristic energy surface from which the excitation energy is measured.

The cross sections for (p, n) reactions were measured by the activation technique in the projectile energy range 5 to 10.5 MeV for target nuclides vanadium-51, chromium-52, copper-63, copper-65, silver-107, silver-109, cadmium-111, cadmium-114, and lanthanum-139 (73). Approximate proton reaction cross sections were obtained below the $(p, 2n)$ threshold by adding to the (p, n) cross sections the previously reported charged-particle emission (p, q) cross sections. These experimental results are compared at a few energies with volume absorption and surface absorption optical-model calculations for the proton reaction.

Fission fragment angular distributions were measured with gold surface barrier detectors for proton-induced fission of uranium-233, uranium-234, uranium-235, uranium-236, uranium-238, neptunium-237, and plutonium-239. The proton projectile energy range of 4 to 12 MeV was obtained with the Argonne tandem Van de Graaff (77). Fission cross sections were determined for targets of uranium-233 and uranium-238. Over this energy range, the fission cross section of uranium-233 changed by a factor exceeding a million. Total reaction cross sections were calculated for the highly fissionable uranium-233 target by applying small corrections (3 to

20%) for the spallation cross sections. These reaction cross sections are compared with theoretical values deduced from surface absorption optical-model calculations (77), and excellent agreement is obtained.

Excitation functions of the ground state (0^+), the 1.43-MeV first excited state (2^+), and the 2.37-MeV second-excited state (4^+) in chromium-52 were measured for the ^{55}Mn (p, α)^{52}Cr reaction (81). The differential cross sections were measured at 90, 130, 150, and 170 degrees in 5-keV steps with protons accelerated by the Argonne tandem Van de Graaff in the energy range 9.5 to 9.9 MeV. The energy loss of the beam due to the target is less than 2 keV. The energy resolution of the proton beam is less than 5 keV. The reaction particles were identified with surface-barrier solid-state detectors, which subtended an angle of 7 degrees in the reaction plane. In order to discriminate against protons, the detectors were biased to just stop the ground-state alpha particles.

A typical excitation function is displayed in figure 15A. Estimates of the ratio of the level width to spacing, Γ/D, in the compound nucleus iron-56 exceeds 50 at our excitation energies of about 20 MeV. Hence, we analyzed our excitation functions in terms of the statistical fluctuations of Ericson (Ann. Phys. [N.Y.] $\underline{23}$ 390 [1965]). The autocorrelation function defined by,

$$R(\varepsilon) = \left\langle \; [\sigma(E) - \left\langle \; \sigma \; \right\rangle][\sigma(E+\varepsilon) - \left\langle \; \sigma \; \right\rangle] \; \right\rangle / \left\langle \; \sigma \; \right\rangle^2 \qquad [14]$$

was calculated for each of the 12 excitation functions. Comparisons of these autocorrelation functions with the corresponding theoretical expression,

$$R(\varepsilon) = R(0)\Gamma^2/(\Gamma^2 + \varepsilon^2) \qquad [15]$$

gave values of the two characteristic parameters of the Ericson theory, R(0) and Γ, where R(0) is equal to

$$R(0) = (\left\langle \; \sigma^2 \; \right\rangle - \left\langle \; \sigma \; \right\rangle^2)/\left\langle \; \sigma \; \right\rangle^2 \qquad [16]$$

and Γ is the characteristic width of the compound states. The data in figure 15B represent an average of 12 autocorrelation functions (three states and four angles). The data were normalized to unity at $\varepsilon = 0$ before averaging. The fluctuations at large ε are considerably damped in the average correlation function as expected.

The effect of the experimental resolution on the parameters R(0) and Γ was investigated. Resolutions of 10, 15, 20, and 30 keV were simulated by averaging 2, 3, 4, and 6 of the original cross-section values and recalculating R(0) and Γ for all excitation functions. From an extrapolation of these data to zero energy resolution, the value of R(0) increases 15 percent and the value of Γ decreases by 20 percent in order to correct for the resolution effect. Hence, we conclude that the width of the compound states at this excitation energy is 5 ± 1 keV.

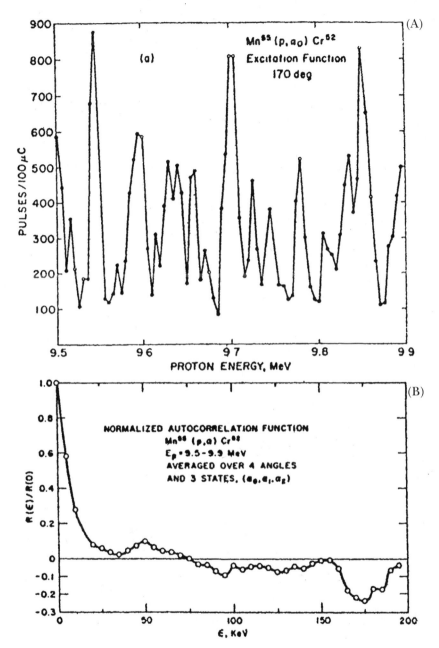

Figure 15. (A) Excitation function for the ^{55}Mn (p, α_o)^{52}Cr reaction (81). (B) Normalized autocorrelation function where this function is the average of 12 normalized [R (0)=1] autocorrelation functions derived from the excitation functions for three states (each at four angles) in the ^{55}Mn (p, α)^{52}Cr reaction (81).

D. New Nuclides and Decay Schemes

The existence of an extinct natural radioactivity has considerable significance in the fields of geochemistry and cosmochemistry. An extinct natural radionuclide is one whose lifetime is too short for detectable amounts to be present from the time of nucleogenesis, yet long enough, to produce through radioactive decay, effects in nature that may be identified at present. A radionuclide whose half-life is in the range from 3×10^7 to 3×10^8 years would fall into this class.

The possible importance of lead-202 and lead-205 in cosmological problems led us to search for these isotopes by long deuteron bombardments of thallium (27). The lead-202 was expected to be produced by the (d, 3n) reaction on thallium-203 and the lead-205 by the (d, 2n) reaction on thallium-205. After the bombardment, lead was chemically separated from thallium and analyzed in a 12-inch, 60-degree mass spectrometer with a multiple filament surface ionization source. Lead-202 was positively identified, while lead-205 was uncertain in that about 80 percent of the mass 205 peak was due to thallium-205. Lead-202 was determined to decay by L and K electron capture with a half-life of 3×10^5 years.

In a further attempt to positively identify lead-205, lead was bombarded for a long period with 21-MeV deuterons in the Argonne cyclotron (35). The bismuth activities were chemically separated and absorbed on a Dowex anion resin column. Lead was eluted from the column with 0.1 normal hydrochloric acid many times during the first three days to remove all lead contamination. After a growth period of two weeks, in which bismuth-205 decays to lead-205, the lead was again eluted, given further chemical purification and mounted for counting. Lead-203, decay product of bismuth-203 served as a tracer, and was given time to decay before the lead-205 counting began.

The lead-205 radiations were examined with a gamma scintillation spectrometer employing a one-eighth inch sodium iodide crystal with a beryllium window. The gamma spectrum associated with the electron capture decay of lead-205 (1.8 counts per minute) was shown to be identical to the L x-ray spectrum of lead measured from a mixture of electron-capturing bismuth isotopes. From these data and some assumptions about cross sections and yields, we estimated the L-electron capture half-life of lead-205 to be approximately 5×10^7 years (35). In another study (39), a lower limit of 10^{10} years was determined for the partial half-life of lead-205 for K-electron capture.

In view of the possibility that extremely old lead ores contain detectable quantities of radiogenic thallium-205 and the interesting application of such measurements to the time interval between the formation of elements and the deposition of ores, we have made additional measurements of the lead-205 L-electron capture half-life (50).

Two lead samples enriched to 26.6 percent in lead-204 were irradiated in the MTR. The first sample containing 100 milligrams of lead oxide received a total integrated flux of 1.7×10^{21} neutrons per square centimeter over a period of three months. The second sample of 150 milligrams of lead oxide received a total integrated flux of 7.51×10^{21} neutrons per square centimeter over a period of one year. The lead-205/lead-204 ratios of the samples from the two above irradiations are 0.001175 ± 0.000060 and 0.005064 ± 0.000063, respectively, as determined by mass spectrometry. From these experiments, the partial half-life of lead-205 for L-electron capture is $(3.0 \pm 0.5 \times 10^7$ years). The K/L electron capture ratio of lead-205 is $\leq 6 \times 10^{-4}$, and the disintegration energy is probably less than 86 keV. No gamma rays were observed in the decay of lead-205.

Approximately 10^5 counts per minute of titanium-44 were produced by the (d, 3n) reaction on a target of scandium-45 (41). Some three weeks after the bombardment, the titanium was chemically separated from the scandium. The titanium fraction containing the titanium-44 was examined with a gamma scintillation spectrometer equipped with sodium iodide crystals of one-eighth and two-and-one-half inch thicknesses. The titanium-44 decays by electron capture predominately to a 144-keV level in scandium-44, which in turn de-excites by two gamma rays in cascade.

In a further search for extinct radionuclides, several nuclides were exposed for a one-year irradiation in the MTR. One of these samples was a 300-milligram sample of HfO_2, enriched in hafnium-180 to 93 percent. Three years after the completion of the irradiation, when the activities of hafnium-175 and hafnium-181 had decayed by many thousands, the sample was dissolved and chemically purified (64).

A fraction of the sample was isotopically analyzed in a 12-inch, 60-degree mass spectrometer with a multiple-rhenium-filament ionization source. The new hafnium isotope, hafnium-182, was detected in a hafnium-182/hafnium-180 ratio of 0.00147 ± 0.00001. This new isotope of hafnium decays with a half-life of $(9 \pm 2) \times 10^6$ years by beta emission predominately to a 271-keV level in tantalum-182. The number of 271-keV gamma rays per beta disintegration is 0.84 ± 0.10. The log ft for the beta transition to the ground state of tantulum-182 is greater than 15 indicating that this transition is at least third forbidden.

It had been known for some years that the favored alpha-decay transitions in odd-A nuclei systematically populate the rotational band for which the orbit of the last odd nucleon remains unchanged. This well-tested rule seemed, however, to be violated in the alpha decay of plutonium-239, with a measured ground-state spin of ½+, where the favored alpha-decay transition was reported to populate the ground state of uranium-235, with a measured spin of 7/2+. To have such an ambiguity in the decay between two of the most important nuclei in the Atomic Energy Commission's program was an intolerable situation. I first thought a mistake existed in one of the two spin values, but consulting with the experts, this didn't seem probable. I then thought that uranium-235 must have an isomeric state of very low energy with a spin of ½+, which up to that time had gone undetected. Hence, I began a search for an isomeric uranium-235 with a spin of ½+ (42).

In the first experiment we collected uranium-235, freshly milked from the plutonium-239. An upper limit of four months for the half-life of an I= ½+ excited state of uranium-235 was deduced from a spin measurement of this uranium-235. Next, counting experiments with scintillation and proportional spectrometers eliminated half-lives of less than or equal to four months for possible isomeric transitions in uranium-235 with sufficient energy to convert in the M shell. We concluded, therefore, that gamma rays following the 5.150-MeV alpha particles were converting in the N shell or beyond with a half-life of less than four months.

In the next series of experiments, uranium-235 samples were separated in a few minutes from plutonium-239 by (a) an ether extraction of uranium from a plutonium-239 solution reduced with Fe^{++} and saturated with ammonium nitrate, (b) equilibration of the ether with several similar washes, and (c) evaporation of the ether containing the uranium onto a platinum plate.

The samples were counted in a Bradley PC-11 proportional counter with a loop wire electrode. Least-squares analyses of decay curves gave a half-life of 26 minutes for the uranium-235 isomer (42). The counting rate in the Bradley counter increased sharply with applied voltage, a phenomenon characteristic of soft radiations. Absorption experiments with uranium-235m samples prepared by a recoil technique onto thin carbon films served to show that the conversion electrons were of extremely low energies.

The 1.1-hour lead-204m was prepared by the (d, 4n) reaction by bombarding lead-206 with 22-MeV deuterons from the Argonne cyclotron. The anisotropies of the three different pairs of gamma rays in the triple cascade of the decay of 1.1-hour lead-204m were measured (36). From these results, the multipolarities of the three gamma rays were determined.

Neptunium isotopes of mass numbers 234, 235, and 236 were produced by an 18-MeV deuteron bombardment of uranium highly enriched in uranium-235 (46). The radiations of neptunium-235 were studied with scintillation and proportional counters in singles and coincidence. The L/K and the M/L electron-capture-branching ratios of neptunium-235 were measured to be 36.7 and 0.46, respectively. From the L/K ratio, the calculated decay energy of neptunium-235 is 123 keV. Most of the electron capture goes to the ground state of uranium-235. The alpha-particle to electron-capture decay ratio for neptunium-235 is 1.59×10^{-5}, giving a partial alpha-particle half-life of 7.0×10^4 years. The experimental data favors a spin of 5/2 for neptunium-235.

Fourteen months after the above deuteron bombardment of uranium, a portion of the neptunium fraction was analyzed in a 12-inch 60-degree mass spectrometer using a multiple filament ionization source. The amounts of mass numbers 234, 235, 236, and 237 present at the time of analysis were < 0.25, 72.7, 25.0, and 2.3 mole percent, respectively. The long-lived isomer of neptunium-236 was present in sizable amounts and positively identified (46).

The absolute alpha-particle energy of the polonium-210 alpha particles was measured by a 180-degree magnetic spectrometer (47). The value obtained was

5.3054 ± 0.0010 MeV. Measurement of the energy (α_0) of the alpha-particles from curium-244 leading to the ground state of plutonium-240 was made relative to the energy of the polonium-210 alpha particles. A value of 5.8025 ± 0.002 MeV was obtained for this curium-244 α_0 energy. The energy difference between the ground state transition α_0 and the transition α_1 to the first excited state of plutonium from curium-244 was measured to be 43.5 ± 1 keV.

Measurements of K and L conversion coefficients of M1 transitions in nuclei with mass numbers around 200 indicate values significantly lower then theoretical results computed by Rose, in the point nucleus approximation. Hence, a study (during my Fulbright year in Amsterdam) was made of the conversion coefficients of the 270- and 404-keV transitions in thallium-203 (53). For the 279-keV transition, α_k=0.163 ± 0.003 and α_L=0.0487 ± 0.0012. For the 404-keV transition, α_k=0.118 ± 0.011.

We made the first determination of the spontaneous fission half-life of curium-246 from a plutonium fraction, which had been chemically separated from uranium and transplutonium elements a week after a thermonuclear explosion (37). The chemically separated plutonium fraction (containing [4.7 ± 0.8] x10^{-12} grams of plutonium-246) was left undisturbed for sixteen days, after which time a chemical separation of the curium was made (the half-lives of plutonium-246 and americium-246 are 11.2 days and 25 minutes, respectively). The chemical yield of the curium-246 was determined by use of a chemical tracer. After fission counting this sample for over a hundred days, a spontaneous fission half-life of (2.0 ± 0.8) x10^7 years was obtained for curium-246 (37).

E. Isomeric Cross-Section Ratios

A research program was initiated to study the cross sections of low- and high-spin isomers in (n,γ) and (γ, n) reactions. The aim of this work was to show the degree to which the isomeric cross section ratio can give information about the dependence of the energy-dependent-level density on spin and the spins of the initial compound states formed in a nuclear reaction (60). We constructed a detailed treatment of the de-excitation process in order to relate isomeric cross-section ratios quantitatively to the spin dependence of the nuclear level density and the multiplicity and multipole character of the gamma-ray cascade.

Among the important factors that determine the isomeric ratio are (a) the spins of the compound nuclear states, (b) the number and types of steps in the de-excitation of the compound state; this depends on the excitation energy, (c) the angular momentum carried away at each step, (d) the probability of forming states of different spins during each step of the cascade, and finally, (e) the spins of the isomeric states.

To compare experimental and theoretical results for neutron-capture reactions, I had to make more specific assumptions in terms of the general considerations outlined above. (a) It is assumed that for thermal or resonant energy neutrons only s-wave neutrons are captured, so that the spin of the capturing state in the compound nucleus is given by $I \pm 1/2$, where I is the spin of the target nucleus. (b) The average number of steps in the gamma-ray cascade (the gamma-ray multiplicity) has been measured for quite a few nuclei and is approximately 3 to 4. The multiplicity was kept as a variable in the calculations, although the calculations are in general consistent with the measured multiplicity. (c) The gamma-ray cascade is believed to consist mostly of dipole radiation. Levels of both parities are assumed to be present in equal numbers, so that parity changes are not followed in the cascade process. The parity of the initial compound state and of the final isomeric states might become important if one makes the restrictive assumption of electric dipole radiation. However, the effect is largely washed out if the distribution about the average number of gamma rays per cascade is broad enough so that there are approximately equal numbers of cascades with even and odd numbers of transitions. (d) Of the different factors which have to be taken into account, the relative probability of forming states of different spins is the most model dependent. The total radiation width for emission of dipole radiation from a state of spin J_c and initial excitation energy B can be written as:

$$\Gamma_\gamma = C \int_0^B f(E,B,J_c,J_f) E^3 \frac{\rho'(B-E)}{\rho(J_c,B)} dE, \qquad [17]$$

where C is a constant, $f(E, B, J_c, J_f)$ is a model dependent factor, E is the transition energy of the radiation,

$$\rho'(B-E) = \sum_{|J_c-1|}^{J_c+1} \rho(J,B-E)$$

is the total density of levels at excitation energy (B-E) which are accessible to the initial state (spin J_c) by emission of dipole radiation, and $\rho(J_c, B)$ is the density of levels of spin J_c at an excitation energy B. In the calculations, we are concerned only with relative probabilities for decaying to different spin states, so the constant C and level density factor $\rho(J_c, B)$ can be ignored. The factor $f(E, B, J_c, J_f)$ is taken as unity in most of our calculations because of the variations in the nuclear matrix elements, which are assumed to have been averaged out by considering a sufficiently large number of initial and final states of the nucleus. The factor $\rho'(B-E)$ contains the spin dependence of the nuclear level density, predicted theoretically to be of the form

$$\rho(J) \propto \rho(0)(2J+1)\exp[-(J+1/2)^2/2\sigma^2], \qquad [18]$$

where $\rho(J)$ is the density of levels with spin J, $\rho(0)$ is the density of levels with spin zero ($\rho(0)$ contains most of the dependence of the nuclear level density on excitation energy), and σ is the parameter that characterizes the distribution in spin.

The relative probability of forming each member of a pair of nuclear states by the (n, γ) and (γ, n) processes has been compared with theoretical predictions described above (60, 71, 85) for a large number of reactions. In general, the agreement is satisfactory. The calculations are sufficiently consistent with experiments to make their predictions useable as a guide for assigning spins to compound states formed in either thermal or resonant energy neutron capture.

In a subsequent paper (61), the calculational procedures described previously were extended to include particle emission before the gamma-ray cascade. The aim of this work was (a) to make detailed and extensive measurements of the isomeric cross-section ratios for the same isomeric pair produced in a large variety of reactions and (b) to explore quantitatively the significance of some of these results in terms of the angular momentum transfer in the initial interaction and the modification of this angular momentum by the sampling of the spin dependence of the nuclear level density by particle and gamma-ray emission. Such analyses of the experimental cross-section ratios yield valuable information on the spin dependence of the nuclear level density, especially if one considers reactions where particles are emitted which carry off enough angular momentum to reach many spin states of the residual nucleus.

The first step in this more general approach is the calculation of the distribution of angular momentum, J_c, of the compound nucleus. This is given by

$$\sigma(J_c, E) = \pi \lambdabar^2 \sum_{S=|I-s|}^{I+s} \sum_{l=|J_c-S|}^{J_c+S} \frac{2J_c+1}{(2s+1)(2I+1)} T_l(E),$$ [19]

where $\lambdabar$ is the de Broglie wavelength of the incoming projectile, s is the spin of the projectile, I is the spin of the target nucleus, and $T_l(E)$ is the transmission coefficient of a particle with orbital angular momentum l and energy E.

After calculating the distribution function for the angular momentum J_c of the compound nucleus, one must examine the modification of this distribution function by emission of the neutrons and finally gamma rays. The relative probability for a compound state with angular momentum, J_c to emit a neutron with orbital angular momentum l leading to a final state with angular momentum J_f is given by

$$P(J_f) \propto \rho(J_f) \sum_{S=|J_f-\frac{1}{2}|}^{J_f+\frac{1}{2}} \sum_{l=|J_c-S|}^{J_c+S} T_l(E),$$ [20]

where $T_l(E)$ is the transmission coefficient for a neutron with orbital angular momentum l and energy E.

The probability of populating a final state with spin J_f by particle emission from a compound state of spin J_c, given by Equation [20], depends on the level density $\rho(J_f)$, which is given by Equation [18]. It was found that using the average neutron energy was an accurate approximation. If a second neutron is to be evaporated, the process is repeated. After neutron emission is energetically forbidden, the gamma-ray cascade described earlier is followed.

As an example of a pair of isomers formed in many different reactions, we consider the mercury-197 isomers (61), mercury-197 (I=1/2), and mercury-197m (I=13/2). These isomers were produced in the following reactions: (a) gold-197 (p, n) with proton energies 7.3 to 10.4 MeV, (b) gold-197 (d, 2n) with deuteron energies 7.2 to 21.4 MeV, (c) platinum (α, xn) with helium-ion energies 18.4 to 27.3 MeV, (d) mercury (n, γ) at thermal energies, (e) mercury (n, 2n) with 14-MeV neutrons, (f) mercury-196 (d, p) with 11-MeV deuterons, and (g) mercury (α, α'n) with 41-MeV helium-ions.

In deciding whether the last gamma-ray transition populates the I=1/2 or 13/2 state, one must consider the I=5/2 state lying between the ground and isomeric state. It is assumed that states with $I \geq 11/2$ populate the isomeric state, states with $I \leq 7/2$ (directly or through the I=5/2 states) populate the ground state, and states with I=9/2 populate both the ground and isomeric states.

Excitation functions for the gold-197 (d, 2n) reaction are shown in figure 16A. One sees that the excitation function for the high-spin isomer is displaced to higher energies, illustrating the effect of the initial angular momentum on isomer yields. In figure 16B, the statistical-model calculations described above are applied to the calculation of the yields of the mercury-197 isomers produced by the gold-197 (d, 2n) reaction. By comparison of the theoretical and experimental results shown in figure 16B, one concludes that the parameter σ which characterizes the dependence of the nuclear level density on angular momentum has a value of 4±1. This result is consistent with that obtained for all the other reactions where compound nucleus formation predominates.

Two important conclusions can be drawn from the comparison of these calculations with all the different experimental results. In the first place, the same value of σ is obtained irrespective of whether the reaction leading to the products is a (p, n), (d, 2n), or (n, 2n). This provides evidence in support of the compound nucleus model, if one characterizes the compound nucleus by its angular momentum J_c as well as by its excitation energy. Second, analyses of isomeric cross-section ratios can provide information about the spin dependence of the level density.

In contrast, direct reactions lead to different results. Relatively small amounts of angular momentum are transferred in reactions that proceed predominantly by a direct-interaction mechanism. Such reactions, therefore, give a larger yield of the isomer with spin closer to that of the target nucleus.

Excitation functions for producing the indium-110 isomers (I=2 and 7) were measured for the silver-107 (α, n) and silver-109 (α, 3n) reactions with isotopically separated targets (80). Nuclear temperature and isomer ratios were calculated also on the basis of a semi-quantitative theory for a Fermi gas with pairing correlations analogous to those of a superconducting metal (84). In general, the degree of agreement between the superconductor theory and the various experimental isomer data for all types of nuclei is unimproved over the simpler shifted Fermi gas theory.

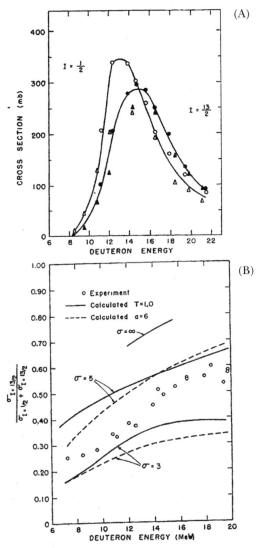

Figure 16. (A) Excitation functions for the gold-197 (d, 2n) mercury-197 (open symbols) and gold-197 (d, 2n) mercury-197m (solid symbols) reactions as a function of deuteron energy. The circles and triangles refer to two separate runs (61). (B) Comparison of the experimental (open circles) and calculated isomer cross-section ratios σ (I=13/2) / [σ (I=1/2) + σ (I=13/2)] for the gold-197 (d, 2n) reaction. The solid line curves were calculated for two values of the spin-dependent parameter σ and a constant nuclear temperature of one MeV for the emitted neutrons. The dashed curves were calculated assuming the nuclear temperature for the emitted neutrons varied with excitation energy as predicted by the degenerate Fermi gas model (61).

The isomer ratios were calculated with a statistical model computer program. This program is divided into three main parts: (a) the computation of the partial compound nucleus cross sections and the normalized initial compound nucleus spin distribution, (b) the computation of the energy and normalized spin distribution following particle emission, and (c) the computation of the energy and spin distribution following gamma-ray emission. This program was published as Argonne National Laboratory Report ANL-6662 in 1962. There were many requests for this program and it became widely used in many countries.

The University of Chicago was the contracting agent for the AEC in the operation of the Argonne National Laboratory. It was through the university that Argonne staff participated in the retirement benefits of TIAA (Teachers Insurance and Annuity Association). For some time there had been under discussion means for strengthening the interaction between scientists in the departments of the university and the divisions of the Argonne National Laboratory. As a result, the university established a limited number of joint appointments which were to be offered to members of the Argonne staff who were interested in taking part in instruction on the campus. In March, 1963, I was one of two Argonne scientists offered such an appointment, with the responsibility of teaching a course during one of the quarter sessions, supervising graduate students and taking part in the seminars, and other scientific activities of the Chemistry Department on campus.

For some time, I had been campaigning for a heavy-ion accelerator at Argonne, to be used by the surrounding university scientists as a national facility. As early as April 4, 1960, I prepared a list of possible users from universities in the Midwest. Again in December 2, 1963, I wrote a letter to the Laboratory Director's office proposing a variable energy cyclotron capable of accelerating heavy ions. I stated that, "the production and study of new heavy and superheavy nuclei require a variable energy cyclotron capable of accelerating intense beams of approximately 1 milliampere of a wide variety of heavy ions." I then went on to list some of the required characteristics of such a heavy-ion accelerator.

My efforts were successful to the extent that a proposal was developed at the Laboratory and submitted to the AEC for funding. However, in order to get the proposal to have the backing of the Physics Division, the proposed accelerator required, in addition to the heavy-ion capability I sought, an elaborative light-ion capability. Hence, the final design of the proposed machine became very complex and, as a result, very expensive to build. Funding was never approved.

My request for a stand-alone and versatile heavy-ion accelerator was ahead of its time. Years later, the Lawrence Berkeley Laboratory's Hilac was rebuilt (SuperHilac) to accelerate very heavy ions and eventually a variable-energy heavy-ion cyclotron was built at Michigan State to serve as a national facility.

V. First Guggenheim Year (1964–1965)

On March 17, 1964, I was informed by Dr. Gordon N. Ray, president of the John Simon Guggenheim Memorial Foundation, of my appointment as a Guggenheim Fellow. In April, I received notice from the French National Minister of Education of my appointment as a visiting professor at the University of Paris for the 1964–65 academic year. This academic appointment made me an employee of the French Government with the very important benefit of national health care for my entire family.

Anticipating that we would be on sabbatical in Paris during the next school year, we began right away to make contacts about schools in Paris where our children might possibly enroll. Since there are always numerous American families in Paris, especially military families, American schools have been established in the Paris suburbs to accommodate those who choose to live in an American enclave. Through others who had lived in Paris, we learned about the Lycée de Sevres, Sevres being a suburb in the south of Paris. This Lycée had an International Lycée as part of its complex, where the lessons and textbooks were in French; however, the teachers were knowledgeable in English. The students in the international section were mostly from families of diplomatic personnel serving in Paris.

We chose to send our children to the International Lycée de Sevres; however, there was an age restriction of 12 years. Hence, only Linda (16), Jann (14), and Robert (12) were eligible, and we had to make other arrangements for Joel (10). After entrance examinations, the older three children were accepted and we completed the entrance arrangements later in July when we arrived in Paris. We enrolled Joel in a French school.

The period leading up to my departure was a very busy one. I had a number of manuscripts, which had to be completed and submitted to journals. In November, 1963, I received an invitation to be a lecturer at the Fifth Scottish Universities' Summer School in Physics, to be held from July 26 to August 15, 1964. The location of the School was in Newbattle Abbey College, Dalkeith, near Edinburgh. Professor Dee from the University of Glasgow and Professor Feather from the University of Edinburgh were joint directors of the school. I was asked to submit an abstract of my lectures and a reading list some months before the opening of the school. In addition, the lectures at the school were to be published, and I had to prepare a manuscript to be available, before the school opened, for duplication to be given to the school's participants. The participants, coming mostly from Europe and the United States, were in large part established scientists.

Some months before leaving for Paris, my wife learned of a public television station offering an early weekday morning program on the French language and culture. The whole family gathered at 7 AM for breakfast and listened faithfully to the program over some six months. We had not had a television set until we purchased one for this program. The program was very helpful in offering each morning a window into a small segment of French life. For example, one morning it gave a picture of the life of a Paris taxi driver. During our first taxi ride on entering Paris, Robert exclaimed, "it's just like the TV program," confirming that the television program had depicted well the real-life situation.

During this sabbatical, we decided not to have a large American-model car in Europe. We ordered a French-model Peugeot-404 station wagon from a dealer in the Chicago area, some months before sailing, to be delivered in Paris. We were able to rent our furnished home in Western Springs, a suburb southwest of Chicago, to a family spending a year at Argonne. We were traveling with limited luggage, so some of our family supplies and clothes were shipped ahead to be delivered in Paris. In early July, 1964, friends drove us to O'Hare Airport in Chicago and we flew to New York. We spent a couple of days in New York, taking in the New York World's Fair and some sightseeing. In New York harbor, we boarded the French-line ship, the S. S. *France,* and sailed for Le Havre, France. On our previous sabbatical we sailed on two ships of the Holland American line to and from Amsterdam. The S. S. *France* was more luxurious with first-class service and food. On arriving in Le Havre, we boarded a train for Paris and then the taxi to our hotel.

We spent a few days in Paris, picking up our car, finalizing school arrangements, and searching for a suitable apartment. We had contacted the American Embassy in Paris early on for rental listings of apartments. After investigating several places, we settled on a furnished and comfortable three-bedroom apartment in a multi-apartment building, with a resident concierge, in Saint Cloud. Saint Cloud is a suburb south and on the edge of Paris adjacent to Sevres. The metro service comes near, and the children had train service to the Lycée de Sevres. Saint Cloud is on the direct route from Paris to Versailles. Although expensive, the apartment on the third floor was attractive and served our needs well.

After initial busy days in Paris, we departed Paris by car for the Scottish Summer School near Edinburgh. Our Peugeot had three rows of seats, the back row facing the rear; however, the seating space was adequate for six people. The problem was that there was no additional space for luggage. This was solved by a roof rack and a canvas covering. Each person was allotted one-half of a large suitcase and, in addition, the women each had a small make-up case. Our plan was to travel some six weeks, in this mode, throughout Europe up to the time school began.

After crossing the channel to England, we stopped in London for sightseeing including Parliament, Westminster Abbey, Madame Tussaud's Wax Museum, the London parks, and more. We proceeded north stopping at Anne Hathaway's cottage in Stratford-on-Avon. We made a point of locating the well-known English "Bed and Breakfast" for lodgings on the way to Scotland.

The stay at Newbattle Abbey College was enjoyable for the entire family. The summer school was conducted in a very informal manner and there were many opportunities for questions and discussions. My series of lectures at the school were entitled "Structure of the Transition State Nucleus in Nuclear Fission." All of the lectures were published in the book *Nuclear Structure and Electromagnetic Interactions,* published by Oliver and Boyd, Edinburgh and London (1965). My lectures appear on pages 319–374 (87).

These printed lectures continue to give a very exhaustive treatment of nuclear fission near threshold. The introduction describes the expected low-lying channels of even-even and odd-A transition-state nuclei. From the excitation energy dependence of the fission fragment angular distributions, it is possible to characterize the saddle states in terms of energy and quantum numbers I and K. The extended axis of the transition-state nucleus is assumed to be identical in direction of the two fragments after separation. The transition-state nucleus is endowed with an angular momentum K about the extended axis, which arises from the angular momentum of collective rotation, and the angular momentum of unpaired nucleons about the axis of extension. For an even-even transition-state nucleus the angular momentum K for the low-lying states arises only from collective rotation about the axis of rotation. The dynamics of the fissioning nucleus can be compared with the dynamics of a symmetric top (see figure 17). The axis of symmetry of the top corresponds to the extended axis at the saddle-point deformation, and after scission, with the axis along which the fission fragments separate. The probability distribution in direction of the fission fragments is assumed to be equal to the probability distribution in direction of the symmetry axis of a symmetric top, which has the same angular momentum quantum numbers I, M, and K defined in figure 10. Conservation of angular momentum requires that I and M are constant throughout the various extended shapes on the path to fission. No such restriction holds for the parameter K. However, evidence shows that once the nucleus is at or past the saddle shape, K is a good quantum number.

Following this extended introduction, the lectures describe then-current information on fission fragment angular distributions as obtained from (n, f), (d, pf), and (α, α'f) reactions. The final lecture describes fission fragment anisotropies at saddle-point excitation energies considerably in excess of threshold. At these energies, information is obtained about the shape of the transition nucleus.

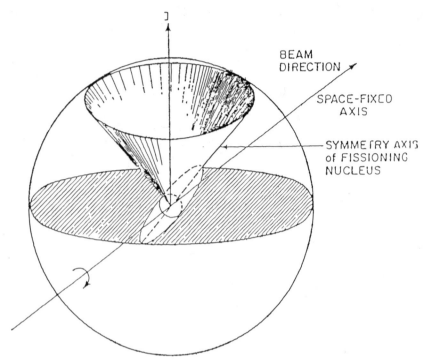

Figure 17. Classical illustration of the probability distribution of fission fragments. The long axis of the prolate spheroid represents the symmetry axis of the fissioning nucleus. For a fixed value of K and the I vector in the diagram, the symmetry axis can lie anywhere on the cone. In addition, if I has equal probability to lie anywhere in the plane perpendicular to the beam direction, the I axis has to rotate by 360 degrees to determine the fission fragment distribution for a unique value of K and I. In an experiment the fission counters are usually placed in the plane of the figure, which contains the beam. The angle θ is measured in this plane from the beam direction (87).

During free afternoons at the school, we traveled in the Scottish Highlands viewing the heather fields and the many sheep roaming the hills on the countryside.

On leaving the Scottish Summer School, my next official meeting was in Hercegnovi, on the Adriatic Sea, in southern Yugoslavia. I had been invited in the spring of 1964 to lecture at the Ninth Summer Meeting of Nuclear Physicists to be held in Hercegnovi from August 23 to September 9, 1964. The timing of these meetings worked out well with my scheduled sabbatical leave in Europe. Our trip to Hercegnovi from Edinburgh brought us through most of Europe. From Scotland we traveled south to Newcastle, England, boarded the S. S. *Lita* and crossed the North Sea to Bergen, Norway. This crossing is an overnight trip.

The North Sea is unusually rough compared with travel on the Atlantic Ocean. All the chairs, tables, and other items were chained down and it was impossible to walk on ship without holding onto a rail.

In Bergen, we went to a Norwegian folk festival with great family entertainment. On the trip from Bergen to Oslo, we experienced several times that the highway would end, necessitating a ferry ride through a beautiful Norwegian fjord. On this trip we stopped at a goat farm in the Norwegian countryside and saw many antelope on the roadside, some coming directly up to the car.

From Norway we crossed into Sweden, which at this time required driving on the opposite side of the road from Norway. We proceeded to Stockholm and made visits to some of the tourist attractions. We next made our way to Denmark and spent some time in Copenhagen, taking in the changing of the guards at the palace. From Denmark we traveled to Germany, spending time in Berlin. Among the places visited were the zoo and the museums in West Berlin, the Brandenburg Gate, and the Wall between West and East Berlin. The crossing from West to East Berlin was especially interesting at this time. It was an eye-opening event for the children, as the car was meticulously searched both inside and out, as well as the chassis underneath. All English newspapers and magazines were confiscated. East Berlin in 1964 was very drab and in need of repair with many unreconstructed buildings. The scenery and small villages on the trip through East Germany, into and out of Berlin, were most interesting. The contrast on crossing the West–East German border was striking. In Bavaria, we visited the former concentration camp at Dachau with its notorious furnaces. Our last stop in Germany was in Munich. From Munich, we made our way to Austria and stopped in Salzburg and Vienna. In Salzburg, we toured the sights including the Castle of Salzburg. The trip across Austria to Vienna passes through beautiful countryside. In Vienna, we visited in the home of a former postdoctoral student who spent two years with me at Argonne and was currently on the faculty of the University of Vienna. Vienna has numerous attractions for tourists with its many small streets peppered with cafés and bakery shops.

From Vienna we traveled south to Yugoslavia passing through Zagreb and Sarajevo, a city that hosted the Winter Olympics some years ago. Along the roadside, we encountered many children and dispensed candy and gum to many appreciative recipients. The roads in Yugoslavia in the 1960s presented many challenges. In fact the roads over the mountains, passing through Sarajevo, to the Adriatic Coast were nearly impassible over long stretches. In addition to being steep and narrow, there were often fallen trees and debris on the road. Part of this was due to the fact that the roads were under major repair with large and primitive equipment. On top of that, the roads were used to transport sheep and goats, frequently requiring stops and long waiting periods. Meeting another vehicle was especially hazardous. The road space was so limited and the embankment so steep that it required an outside guide, to make sure the outer wheels remained on solid ground in order to prevent a catastrophe. It became a family joke that everyone would volunteer to serve as the outside guide, since this was by far the safest position during the passage of two vehicles. Once we reached the Adriatic Coast, the roads were normal. Our first stop on the coast was Dubrovnik. This is a jewel of a city, especially the old city with its huge walls on the sea. The hotels in the new city, constructed by contractors from the Soviet Union, have all the same flaws as those I experienced in the Soviet Union. For example, our new hotel had leaking plumbing, but it also had a beautiful Olympic-size swimming pool. In the old city, hand-made jewelry was a specialty in shops located on narrow and picturesque streets. The Adriatic Coast of southern Yugoslavia is a top-notch vacation area for wealthy Europeans and we experienced this even during the sixties, as the beaches and hotels were crowded with tourists.

The trip from Dubrovnik to Hercegnovi was a short and scenic one. Hercegnovi is another beautiful city on the Adriatic Sea. The conference in Hercegnovi had well over one hundred participants. In addition to those from several institutes and universities in Yugoslavia, participants traveled from Belgium, Brazil, England, Finland, France, Greece, East and West Germany, Hungary, Italy, Poland, Romania, Sweden, Switzerland, and the United States. The pace of the meeting allowed time for questions and the breaks and meals for one-on-one discussions. I met a number of scientists whom I had known previously only by their publications.

On the first leg of the trip back to Paris, we crossed the Adriatic Sea from southern Yugoslavia to Bari, Italy, a port on the southeast coast, directly east of Naples. The trip across southern Italy brought us through a number of villages, all situated on the tops of mountains. I presume such a location gave the settlers added security against invaders. We stopped and examined the ruins of Pompeii, and took in the sights of the Mediterranean Sea as we viewed the remains of Mt. Vesuvius. After a short stop in Naples, we proceeded to Rome, where we spent some time taking in the sights, such as St. Peter's Cathedral, the Vatican, the Pantheon, the Fountain and 3 Coins, the Colosseum, the Swiss guards in Vatican City, and the Catacombs outside Rome. We traveled up the West coast of Italy to Pisa, where we were intrigued by the Leaning Tower. Time was slipping by as we spent September 11 at the Palace of Prince Rainier and Princess Grace in Monaco, as well as the Casino in Monte Carlo. The views of the Mediterranean Sea are spectacular from the elevated drives near the Italian-French border. Located at the foot of the Maritime Alps, the coastline is stunning. Sheer cliffs plunge into the foaming sea, yachts worth millions of dollars cluster in the harbor, and flaming bougainvillea pierce a blue sky. Sparkling lakes and medieval villages make their appearance at every turn. The Grand Corniche Drive, high above the shores of the Riviera, is among the world's greatest drives, and follows an old Roman route built by Napoleon.

From the Riviera, we traveled north from Marseille to Avignon, an ancient city steeped in history. The streets are narrow inside fortified walls. The city has a number of museums and historic sites, and was a major Roman town until the fifth century. Through the medieval period, Avignon had a strong and well-documented history, including the Popes who ruled from there. Some of the sites we visited were the Roman amphitheater, Le Palais de Popes, and the Le Pont d'Avignon. From Avignon, we traveled north, passing through Lyon on our direct route back to Paris. The seven-week family excursion through Europe had come to an end. It was a very enjoyable and highly educational experience for all of us, long remembered for the many great times.

On returning to Paris, we learned that our furnished apartment was not ready for occupancy so we spent seven days in a small hotel. School was to start immediately, requiring a metro and a train ride to the Lycée de Sevres. Joel was enrolled in a French school in St. Cloud, the suburb in which our apartment was located. Joel started in first grade and as he progressed with the language, rose grade by grade, reaching his normal fifth grade by Christmas. To ease the transition, Joel found a good friend his age in his school who was bilingual, the son of an American mother and a French father. The children used the metro to travel all over Paris, a very safe transportation mode at that time.

Immediately I began to work full time at the university. The University of Paris has a long and rich tradition in nuclear science. As an introduction to the laboratory, I will give a very short description of its history and the prominent role played by the Curie family. In 1896, H. Becquerel discovered natural radioactivity. The emitted particles were later identified to be alpha particles (helium-4 ions), beta particles (electrons), and electromagnetic radiation (photons). In 1891, Marie Sklodowska from Warsaw, Poland, went to Paris to continue her scientific studies. She met Pierre Curie, a professor of physics, in 1894 and they were married the next year. Thus began a fruitful scientific collaboration. The Curies and their co-workers, through chemical decomposition and fractionation of pitchblende, an ore rich in uranium and its daughters, discovered almost immediately the element polonium (named after Marie's native Poland). The Curie's work constituted the first exercise in radiochemistry. Starting with two tons of pitchblende, they found radium in a chemically separated barium fraction. The Curies learned that radium could be concentrated from barium by repeated fractional crystallization of the chlorides. They isolated 100 milligrams of radium chloride spectroscopically from barium and assigned the atomic weight of radium to be 225.

In 1903, one-half of the Nobel Prize in physics was given to Becquerel for his discovery of spontaneous radioactivity and one-half was awarded to Marie and Pierre Curie for their discoveries of polonium and radium. The Curies had two daughters, Eve (the author of the well-known book about her mother) and Irène, born in 1897, and who later became a distinguished scientist like her parents. Pierre Curie was killed in a street accident in Paris in 1906 when he walked out in front of a horse and wagon. Marie was appointed to fill his professorship and in 1911 she received the Nobel Prize in Chemistry "in recognition of her services to the advancement of chemistry by discovery of radium and polonium."

Marie Curie had the distinction of being the recipient of the Nobel Prize both in physics (1903) and in chemistry (1911). She died in 1934.

Irène Curie carried on the scientific tradition of the Curies into the second generation. Irène's academic advisor, in addition of course to her mother, was Paul Langevin, a colleague of the Curies and another well-known name in French science. It so happened that Langevin was also the mentor at that time of another bright student named Frederick Joliot. Irène and Joliot later married and both changed their surnames to Joliot-Curie. Together they received the Nobel Prize in physics in 1935 for their discovery of artificial radioactivity. After World War II, Frederick Joliot-Curie became the architect of France's nuclear power program, which now produces eighty percent of the electric power in France. Irène Joliot-Curie was accidentally exposed to polonium (an element discovered by her parents) when a sealed capsule of the element exploded in her laboratory. She developed leukemia but continued to work. Her final contribution in 1955 was to design and draw up plans for the laboratories of the new campus of the university to be located in Orsay, south of Paris and known as the University of Paris-Orsay. Irène died in 1956 at the age of 58, followed by the death of her husband in 1958, who was also age 58.

I had my office in the Joliot-Curie Laboratory on the Orsay campus. The university had three accelerators on the campus. The third one was a variable-energy sector-focused cyclotron for heavy-ion acceleration, which unfortunately came into operation too late for my use. The laboratory had other large pieces of experimental equipment, including mass spectrometers, a large isotope separator, and an alpha-particle magnetic spectrometer. One of my colleagues in the Joliot-Curie Laboratory was a third generation Curie. Irène and Frederick Joliot-Curie had two children. Hélene, born in 1927, was appointed a professor of physics at the University of Paris-Orsay in 1957, and Pierre born in 1933 was a noted biochemist employed by CNRS, a large French National Laboratory. Hélene's husband was Michael Langevin, a grandson of Paul Langevin. He was a nuclear physicist also employed by the Joliot-Curie Laboratory.

Near the end of September we gained occupancy of our furnished apartment and began life as Parisians. There were many small shops across the street from our apartment building for purchasing milk, bread, meat, fresh fruits, and vegetables. We tended to join the French in daily shopping for fresh merchandise. One of the first activities for Dolly and me was to enroll in the Alliance Francaise, an excellent French language school in the heart of Paris. I went to an early morning class on my way to work. When doing this, I'd take a metro into Paris and the train from Paris to Orsay. The trains were punctual and ran frequently during mornings and afternoons to accommodate the large numbers of students traveling during these times.

As mentioned earlier, I had planned to do some heavy-ion experiments during my sabbatical at the Joliot-Curie Laboratory. However, since the new cyclotron was nearly a year late in accelerating its first beam, I used my time for writing manuscripts, library research, and visiting and lecturing at a number of universities and research institutes in Europe and Asia. During the winter quarter (February 5 to April 2, 1965), I gave a course on Nuclear Fission at the University of Paris-Orsay. In addition to university students and staff from Orsay, a number of scientists enrolled from Saclay, a large French National Laboratory near Orsay specializing in atomic energy.

The first manuscript I worked on dealt with the determination of the nuclear level density at a large excitation energy of approximately 20 MeV (88). The purpose of this paper was to demonstrate that if the width of the compound-nuclear states is known from cross-section-fluctuations measurements and the properties of the exit channels are known from other data, the level density of the compound nucleus could be calculated. The level densities of iron-56 and nickel-60 were determined at excitation energies of approximately 10 and 20 MeV by combining calculations based on the statistical theory of nuclear reactions and experimental measurements of level widths Γ from studies of cross-section fluctuations. The importance of each of the various parameters in the statistical-theory calculation was studied as a function of the spin J of the compound nucleus. Assuming a Fermi-gas model with shell and pairing-energy corrections, the evaluated level-density ratios at the two energies gave a value of the level density parameter a of 6.7 MeV^{-1} for both iron-56 and nickel-60. This value of a is in excellent agreement with those results measured at excitation energies of less than 10 MeV. Hence, within the framework of the Fermi-gas model these results give independent evidence, in addition to the evidence already discussed in the literature, that the Γ values determined from cross-section fluctuation measurements are those characteristic of the compound nucleus.

During my stay in Orsay, I also co-authored a 184-page article (105) on "Nuclear-Fission" published in a book entitled *Nuclear Chemistry* (Vol. II) by Academic Press. This work included sections on the discovery, the nature of the fission process, the transition-state nucleus, scission and post-scission effects, and the radioactive decay of the fission products.

I visited and gave invited lectures at a number of places including universities in Heidelberg, Copenhagen, Boudreaux, Mainz, Tübingen, and Zurich, as well as at research institutes in Istanbul, Athens, Saclay, Karlsruhe, Harwell, and the Rutherford Laboratory in England. In May of 1965, I was also invited to give some lectures at the Weizmann Institute of Sciences in Rehovoth, Israel. The latter visit included tours to many of the historic sites in Israel for Dolly and me.

During the 1964 Christmas vacation, more than 100 trains, filled with children of specific ages, left Paris for chaperoned trips to ski villas in the French, Italian, Austrian, and Swiss Alps. We had enrolled our children in early October into this program, and they each joined a group of French children for a ski trip to various resorts in the Alps. Each of our children left from different train stations in Paris at different times and we managed to see all of them off. As a precaution, Dolly and I decided not to leave for skiing ourselves but to remain close to the telephone in case of an emergency. All went well and all of the children had great times, as well as greatly improving their French. Each child enjoyed Christmas parties with groups of their age and made new friends. On their return, we had a busy schedule picking up each child at a different time from a different train station. It was an unusual experience meeting these trains and seeing new friends separating and some lagging behind with legs and arms in casts.

On weekends, we occasionally made a trip to a French site outside Paris. Of particular interest were the Utah and Omaha beaches in Normandy. The year 1964 was the twentieth anniversary of the United States' troops landing on these steep Normandy beaches. The countryside of Normandy and Brittany with their rolling hills and small French restaurants made for enjoyable visits. Mont St. Michel, in the gulf of St. Malo, is a city built on a rock. Crowning the abbey is a tall spire with a statue of the Archangel Michael. The swing in the low and high tides is fourteen meters near Mont St. Michel. One of the more striking sites in France is the city of Carcassonne, an ancient walled city on the Aude River, which looks the same today as it did in the Middle Ages. It is built on the site of a double-walled Roman city of the AD 400s whose fortifications have been rebuilt and restored. We also visited Dijon, the capital of Burgundy, located in the heart of wine country.

During our children's Easter vacation in the spring of 1965, we traveled south to Spain. Our first stop was Bordeaux, where I gave a seminar at the university. Bordeaux is the center of the wine-shipping trade and one of the chief commercial cities of France. The city has a large natural harbor on the Garonne River. Around Bordeaux are sunny hills covered with vineyards. Grapes are made into white and red wines of Bordeaux, which have been famous since Roman days. In the lower southwest of France, where France and Spain meet, we passed through an area inhabited by the Basques, a distinct people with a language of their own. The Pyrenees form a boundary between France and Spain, and we entered along the western end of the mountains where the city of San Sebastian stands. San Sebastian is a beautiful resort on the Bay of Biscay.

On entering Spain we traveled south through Old Castile to Madrid. The University of Madrid is one of the largest universities in Europe. The city is the railroad and transportation center of Spain. The Madrid Post Office, which stands high over Madrid Plaza, is an example of Moorish architecture. One of Madrid's main attractions is the El Prado Museum where we spent most of a day. Another beautiful structure is the Opera House of Madrid, which is famous throughout the world for its beauty. While in Madrid, we also took in a bullfight at the Plaza de Toros. It is the city bullring of Moorish architectural design and capable of seating a very large crowd. The bullfight is a spectacle, made famous in Spain, and an event most people don't want to see a second time. Some forty miles south of Madrid is Toledo, a medieval city of narrow winding streets and ancient buildings of historic importance, which sits high on a hill. Toledo's Alcazar became renowned in the nineteenth and twentieth centuries as a military academy. Farther south on the banks of the Guadalquivir River, is Córdoba, an ancient Moorish city, which was once an important trading center. The streets of the old city are narrow and crooked. The most remarkable building in the city is the great Cathedral.

Traveling still farther south along the Guadalquivir River, we came to Seville, surrounded by vineyards and orange groves. Seville was the home of the Moors for hundreds of years. One still saw the Moorish influence in the network of small, shaded streets, and in quaint, balconied houses built around courtyards and fountains. Seville is second only to Madrid as a center of education and art and also has a great cathedral. We arrived in Seville on Easter weekend and were unaware of the annual religious festival at that time with people coming from all over Spain. It was the first time that we had trouble finding housing. We had traveled all over Europe without reservations and never had a problem. On arrival in the city we encountered scalpers on street corners auctioning off rooms in private homes and warning us that the hotels were filled. Being skeptical, we learned firsthand by many negative responses that the hotels were indeed filled with Spanish visitors for the weekend festival. Hence, we settled for a large room in a private home. We experienced an unpleasant night. All the family became ill from a Spanish meal that evening except Rob, who chose instead to have an American-style hamburger. We learned later that families sacrificed a bedroom or two in their homes on such occasions to earn a small amount of money. The religious festival, with a very elaborate parade through the city, was attended by thousands of people.

From Seville we drove to Gibraltar, a British colony that occupies a rocky peninsula at the southernmost point of Spain. We crossed the Strait of Gibraltar, a narrow body of water that connects the Mediterranean Sea and the Atlantic Ocean, and landed in Africa. We explored Tangier and the city of Tangier, an important seaport of Morocco. Here we rode a camel and shopped in some of the open markets. We drove south to Casablanca in French Morocco. Casablanca is one of the most important ports on the North African coast. Returning to Gibraltar, after a night in Africa, we then proceeded north along the Mediterranean Sea to Malaga, some sixty-five miles from Gibraltar. It is a major seaport and famous for its mild, even temperature and is a popular resort for invalids. In the older section of Malaga is a 700-year-old Moorish castle. Farther north and forty miles inland, we stopped in Granada, in the foothills of the mountains, another old Moorish center. The buildings in Granada are picturesque, showing an oriental style and range from shabby huts of the poor to the glorious Alhambra Palace.

Traveling north along the Mediterranean, we arrived at Valencia, the third largest city in Spain. Valencia was long occupied by the Moors and still has rows of white houses built in the Moorish style. The University of Valencia is one of the best in Spain. Surrounding the city is a fertile plain where most of the land is used for producing oranges, lemons, and grapes. Farther north on the coast of Spain is the major port city of Barcelona, some seventy-five miles south of France. We re-entered France along the coast, on the east end of the Pyrenees, and traveled to Carcassonne, described earlier as a most interesting walled city to visit. From there we passed through Toulouse and made our way back to Paris.

In early September, 1964, I received an invitation to present a paper at the International Atomic Energy Agency (IAEA) Symposium on the Physics and Chemistry of Fission to be held in Salzburg on March 20–22, 1965. The timing of this meeting worked well with my being in Europe. I made plans to attend and presented a paper "On the Role of the Transition-State Nucleus in Fission" (89). Since a large fraction of the excitation energy of the initial compound nucleus is consumed in deformation energy in passing to the fission saddle point, the transition-state nucleus is thermodynamically "cold." Hence, for low excitation energies where the non-fission degrees of freedom favor the passage of the barrier with only a small kinetic energy, it seems reasonable to postulate that the traversal time of the saddle or the lifetime of the transition-state nucleus is many orders of magnitude longer than the characteristic nuclear time. This leads to the prediction that the highly deformed transition-state nucleus will have properties, including a spectrum of excited states, analogous to those of normal nuclei. In the lecture, I described new information on the excited states of the saddle configuration from fission fragment angular distribution measurements.

In a second paper (90), the fission of the compound nucleus uranium-235 was studied near its fission threshold by means of the uranium-234 (d, pf) reaction. The variation of the angular anisotropy with excitation energy has considerably more structure than that obtained by Lamphere for the same nucleus resulting from fast-neutron bombardment of uranium-234 (see figure 18). At least eight fission channels at the saddle point were observed for the energy region between threshold and two MeV above threshold. Mass yields obtained from correlated fragment energies show no variation of the anisotropy with mass ratio. There is, therefore, no evidence from anisotropy measurements of heavy elements that the properties of the saddle point influence the final mass division. The average total kinetic energy release in fission varies less than one-half of one percent for different saddle-point energies observed.

An interesting phenomenon one observes in France, or in Europe in general, is people's refusal to queue up. Those of you who have skied in Europe no doubt have experienced the lack of a line going to the ski lift. Instead people gather in an almost 180-degree angle in front of the place of entrance. Making your way to the entrance requires almost brute force otherwise other parts of the broad front will continue to have priority regardless of how close you are to the desired place of entrance. This was the same at a large cafeteria at the University of Paris-Orsay. The cafeteria offered two sittings for their excellent noon meal, which was subsidized heavily by the French government. Getting a place at the first sitting was a struggle beyond description. Several times I saw people injured during this open competition for a seat. People ate their main meal at

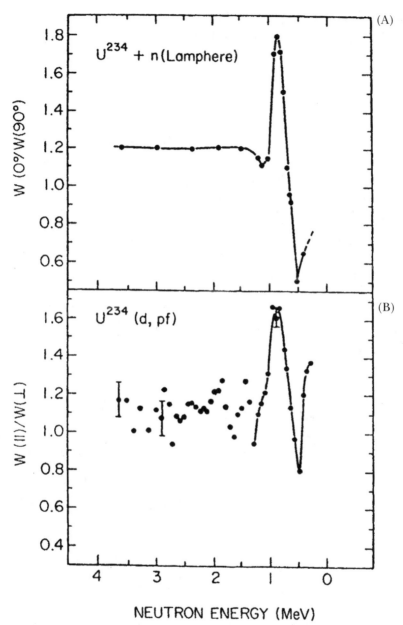

Figure 18. (A) This portion of the figure shows the fission fragment angular anisotropy as a function of neutron energy as obtained by Lamphere (90). (B) This portion of the figure shows the fission fragment anisotropy we obtained in the uranium-234 (d, pf) reaction (90).

noon. This was true at the grade and high schools also. Hence, all of us except Dolly had our large meal at noon for a very small cost. Another custom of the cafeteria that I found unusual was that steak was always available as a back-up meat for those who didn't want the meat of the day.

Many a weekend was spent sightseeing in Paris. We often walked the Avenue des Champs Élysées, the most beautiful avenue in Paris. The Promenade stretches from the Place de Concorde to the Place Charles de Gaulle, the site of the Arc de Triomphe where one has a great view of Paris. This avenue has several theaters and many of the finest shops of Paris. On the Champs Élysées, near the Arc de Triomphe, is a large American-style drug store where people congregate. It has a sizeable fast-food restaurant serving hamburgers, french fries, and milk shakes.

On Bastille Day (July 14), a French national holiday, there was a military parade with all the trimmings on the Avenue des Champs Élysées. We were in Paris during the reign of Charles de Gaulle and his disdain of Americans. He ordered French scientists not to speak in English at international conferences. We went to France knowing de Gaulle's attitude and didn't let it influence us in our relations with the French people. As a joke, our children on visiting Paris, sometimes wore a maple leaf, and in responding to tourists' questions, would in disguise reply in French.

The museums in Paris are well-known around the world, especially the Louvre and the Pompidou center, places we visited many times. Some of the smaller and lesser known museums, but of high quality, are the Rodin Museum, Picasso Museum, and the Museum d'Orsay. The parks in Paris, such as the Gardens of Luxembourg, are well kept and heavily trafficked. The Eiffel Tower, erected for the Paris Exposition of 1889, is an imposing structure attracting many tourists. Linda attended a junior-senior prom on the Eiffel Tower. It is one of the great sights of Paris at night. Walks along the Seine are a good pastime, especially visiting the rows of bookstalls on a Sunday afternoon stroll. On a few occasions, we were able to enjoy a concert at the Notre Dame Cathedral. Trips to the Montmartre and the Sacré Coeur church lead to breathless views of Paris. Often on Sunday we would attend the American Church in Paris, located on the Quai d'Orsay. There were numerous distinguished visitors from the States including astronauts, US senators, and representatives. After church, we would search for a small family restaurant and over the year found many excellent ones. As a tennis enthusiast, I enjoyed the 1965 French Open in the Roland Garros Stadium, where the courts are red clay, giving the tennis ball a long hang-time in the air during a bounce.

In mid-July, after a wonderful and productive year in Paris, we packed up our belongings, shipping some trunks ahead, and drove to Le Havre where we sailed to New York, again on the luxurious S. S. *France*. Arriving in New York, we left immediately in our French Peugeot for the drive back to our home in Western Springs, to resume our various daily schedules.

VI. Argonne Years (1965–1967): Visit to USSR Laboratories

On returning to Argonne, we resumed our experimental programs on the cyclotron, tandem Van de Graaff, and low-energy Van de Graaff accelerators. We immediately prepared our work on the deformation of the transition-state nucleus in energetic fission for publication (92). Fission-fragment anisotropy ratios, W(170 degrees)/W(90 degrees), from fission induced by 43.8-MeV helium ions were reported for sixteen heavy element targets, ranging from the thorium-230 to californium-249. These experimental anisotropy ratios are plotted in figure 19A as a function of the parameter Z^2/A, along with earlier reported values for radium and nuclei near lead. As Z^2/A increases, one sees a sharp decrease in the anisotropy, indicating a marked reduction in the deformation of the transition state nucleus. By modifying values of the fission-to-neutron level width ratio Γ_f/Γ_n, first chance anisotropies were calculated to permit comparison of the targets at nearly uniform and rather high excitation energies. The resulting values of K_o^2 (see Subsection IV.B.2) associated with first-chance fission in conjunction with the assumption of rigid moments of inertia permitted evaluations of saddle deformations (see figure 19B). The saddle deformations were found to be fairly insensitive to programmed variation in Γ_f/Γ_n values based on a semi-empirical relation deduced for the (Z, A) dependence of the compound nucleus. The unrelieved disparity for the heaviest elements between the experimentally derived saddle deformations and those theoretically deduced from the conventional liquid-drop model have been interpreted to suggest a re-evaluation of the fissionability parameter, $(Z^2/A)_{crit}$. Qualitative extrapolation of the new data yields a value of the fissionability parameter of 44 to 45. This lower value compares favorably with results of more recent and unpublished fission theory calculations based on the deformed Nilsson model including nuclear pairing and allowing the surface tension to vary with the curvature of the nuclear surface.

As part of a general investigation of nuclear level densities, the energies of the excited levels of iron-56 were measured by high-resolution magnetic analyses of the particles emitted in the iron-56 (p,p') and cobalt-59 (p,α) reactions (91).

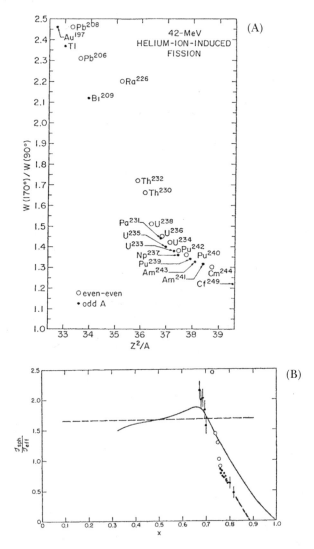

Figure 19. (A) Variation of fission fragment anisotropies for helium-ion-induced fission in various target nuclides (as labeled) as a function of Z^2/A for the compound nucleus (92). (B) Saddle deformations computed from first-chance anisotropies as a function of Z^2/A for the compound nucleus. Solid curve represents theoretical predictions based on charged liquid-drop-model calculations using $(Z^2/A)_{crit} = 50.1$ ($x = 1$). Dashed curve shows extrapolation from protoactinium-231 and heavier elements to zero saddle deformation for a revised estimate of $(Z^2/A)_{crit}$. $\mathscr{I}_{eff}$ is defined by equation [11]. The straight dashed line (nearly parallel to the x axis) represents the theoretical prediction of the critical deformation at the scission point to give a qualitative idea of the difference in saddle and scission shapes (92).

One hundred levels were resolved up to an excitation energy of 6.7 MeV. Two different reactions were used for the determination of the energy levels in order to reduce the probability of missing levels. Data were also taken at two angles.

Evaporation spectra from the two above reactions were investigated with silicon surface-barrier detectors for proton bombarding energies of 9 to 13.5 MeV (94). The level density and the spin cutoff factor σ were calculated from the experimental results as a function of excitation energy in the range of 3 to 9 MeV. The level density varies irregularly with excitation energy at the lower excitation energies and increases smoothly in an exponential manner at the higher excitation energies, reaching some 800 levels/MeV at the highest excitation energy measured.

The nuclear energy levels of cobalt-59 were studied also with a broad-range single-gap magnetic spectrograph. In cobalt-59, an odd A-nucleus, sixty-two levels were resolved in the first 4 MeV of excitation energy (91). Similar measurements were made of the energy levels in chromium-52, manganese-55, and zinc-66 (97). In the first 6.3 MeV of excitation energy, sixty-six levels of chromium-52 were resolved and investigated by the chromium-52 (p,p') and the manganese-55 (p, α) reactions; in the first 3.9 MeV, sixty-two levels of manganese-55 were resolved and investigated by the manganese-55 (p, p') reaction; and seventy-four levels of zinc-66 were resolved and investigated by the zinc-66 (p,p') and gallium-69 (p, α) reactions. In determining the level density as a function of excitation energy for each different nucleus from the experimental data, small corrections are made to the data to correct for energy resolution using the well-known experimental distribution of level spacings (98).

During the Cold War period of the 1960s, the United States and the USSR were exploring the possibility of holding scientific exchanges, in order to foster more friendly relations between the two countries. This program was initiated in December, 1965, at which time a delegation of ten from the USSR, including scientists and state officials, visited the United States. During the trip the USSR delegation visited ten research laboratories in the United States, including national laboratories and universities. In January, 1966, the United States Atomic Energy Commission (AEC) arranged to send a delegation of low-energy nuclear scientists to the USSR consisting of eight scientists, an AEC program officer, and a government official. On short notice, I was notified that I was to be part of the delegation and to prepare the proper official documents for travel to the USSR. Heinz Barschall, Professor of Physics at the University of Wisconsin and Editor of the prestigious journal *The Physical Review*, was the senior member of our delegation.

In anticipation of my travel to the USSR, Dolly and I saw the movie *Doctor Zhivago* (a 1965 film loosely based on a novel by Boris Pasternak) in New York City while attending the winter meeting of the American Physical Society during the period January 26–29, 1966. The winter scenes in this movie were a truly realistic view of what I was to see in my travel through the countryside and an adequate warning of the harshness of the Russian Winters. In addition, everyone prepared me ahead of time for the subzero temperatures in the USSR in February. Hence, I packed warm clothing including a heavy overcoat and lined boots.

We were asked to come to Washington on the evening of February 2, 1966, for a briefing in the State Department on Thursday, February 3. We didn't know until the last minute whether or not the trip would actually take place. Furthermore, we had no details of where we were going, only the promise that the itinerary was going to be decided in Moscow. I met my colleagues in Washington for the briefing, which consisted of some practical advice on travel in the USSR, including a warning about numerous vodka toasts, a tip on how to flush a toilet on an Aeroflot airplane, and the proper procedure for safely cutting into chicken Kiev. It turned out that our menu often did include chicken Kiev, so this advice was helpful in preventing an accident with a hot jet of butter.

We departed from Kennedy Airport at 9:45 AM on February 4 for Paris and stayed overnight in Paris at a hotel on the Left Bank in an area of Paris that was very familiar to me. The next morning we left Le Bourget Airport in Paris on Aeroflot and arrived in Moscow at 9:00 PM on February 5. The temperature on arrival was –40 degrees! (At this temperature one need not specify the temperature scale, because at –40 degrees the Celsius and Fahrenheit scales intersect, –40 degrees Celsius equals –40 degrees Fahrenheit.) We were met at the airport by a group of Soviet officials, scientists, and interpreters and by Mr. Glenn E. Schweitzer of the US Embassy. Among the officials greeting us were Mr. D. P. Filippou and Dr. L. P. Panikov, both of the foreign department of the State Committee, and Dr. A. A. Oglobolin of the Kurchatov Atomic Energy Institute, all three of whom had been in the Soviet delegation to the United States in December, 1965. Also present were two interpreters (we were not totally dependent on these Russian interpreters, however, since one of our scientific delegates was moderately fluent in Russian). With the assistance of the Soviet officials, we were quickly marched through passport control and customs with no baggage inspection. The reception was very cordial and warm, in contrast to the weather, which was blustery and very, very cold. (Some of us were asking at this point whose bright idea it was to send us there in February.) We were taken to the Hotel Sovietskaya, a fancy hotel for VIPs, where we were assigned luxurious rooms. This hotel was at that time in the final stages of a complete reconditioning and is indeed a deluxe hotel; in fact by Soviet standards it's considered an extremely luxurious hotel with high ceilings and wide hallways.

On Sunday, February 6, we toured Moscow with tourist guides. In particular, I remember that we were given a thorough look at the Kremlin, with its intricate rooms and offices. In the evening we were ushered into ten premier seats in the front of a crowded auditorium for a performance of the Bolshoi Ballet. It was obvious to us that Russian citizens were bumped from their seats in order for us to attend this special event. Our delegation was always accompanied by at least one Soviet official of sufficient rank to make things happen on the spot, whether it was obtaining last-minute seats at a sold-out performance, sleeping berths on a sold-out train, or seats on a full plane. Members of our group often referred to this gentleman as our KGB shaman.

On Monday, February 7, we met with members of the USSR State Committee to receive a formal welcome and to discuss the itinerary. Included in the Soviet delegation was Mr. G. Afonin, Chief of the foreign department and Soviet spokesperson. Mr. J. A. Armitage of the US Embassy was also present. At 11:45 AM we went to the Institute of Experimental and Theoretical Physics in Moscow. A lengthy lunchtime banquet was held at the institute. This was the first of many such events, held both at noon and in the evening that we were to experience during our stay. These banquets featured an abundance of food, the menu always including borscht and chicken Kiev, and an endless supply of vodka. The Soviet scientists enjoyed making lengthy and repeated toasts to our friendship, and immediately thereafter gulping the entire content of their vodka glasses. I was sheepishly able to avoid the vodka by substituting my water glass during toasts.

Professor A. I. Alikhanov was the Director of the Institute of Experimental and Theoretical Physics. Among the major pieces of equipment of this institute were a small cyclotron and a heavy water reactor with core flux of 4×10^{13} neutrons per square centimeter per second. This is one of the first (1949) reactors to be operational in the USSR. We spent the afternoon learning about the research programs at these two facilities.

We spent the entire day on Tuesday, February 8, at the P. N. Lebedev Institute in Moscow. This institute has approximately 2,000 employees, including 300 physicists and 300 engineers. They also had 250 students from Moscow University and 75 postgraduate students. This laboratory in 1966 was part of the USSR Academy of Sciences and was one of the better physics institutes in the USSR. It was founded by Peter the Great some 250 years ago. This laboratory has several major experimental facilities including a 260-MeV electron synchrotron, designed and built by Professor V. I. Veksler, and known for its high beam intensity. Although we saw that vacuum tube analyzers were still widely used in USSR laboratories in 1966, we saw at the Lebedev Institute a 100-channel analyzer in operation with transistors.

At 8:45 AM on Wednesday February 9, we left our hotel in Moscow to travel to the Joint Institute for Nuclear Research in Dubna, arriving at 11:45 AM. We traveled in a small bus and the trip was rather slow due to the icy road conditions and heavy traffic. On this trip through the countryside we had an excellent view of the small and primitive wooden homes of the peasants both on the farms and in the small villages. This institute was founded by the Academy of Sciences in 1946 and arose out of the Lebedev Institute. In 1956, the Joint Institute was established to provide research opportunities for scientists of other Socialist countries. Dubna was the Socialist countries' version of CERN, the highly successful laboratory of the European Countries located in Geneva. Dubna consists of five laboratories with some 3,000 employees including 500 scientists. The overall director of the laboratory in 1966 was the well-known and distinguished scientist Professor Bogolyubov. At the time of our visit about one-half of the scientists were from the USSR. The budget of the Joint Institute is shared among participating countries, with about one-half of the funds provided by the USSR and one-quarter by Red China. At the time of our visit, however, the withdrawal of support by Red China created a financial crisis. During my visit I did not see any Chinese scientists, although participants from East Germany, Poland, Korea, North Vietnam, and elsewhere were in evidence.

Of the five separate laboratories at Dubna, the Laboratory of High-Energy Physics is the largest, and at this time its main facility was a 10-GeV proton synchrotron. The Laboratory of Nuclear Problems is the second largest laboratory and operated a 680-MeV synchrocyclotron. The director of this laboratory was Professor Y. P. Dzhelepov whom I had previously met in the home of Professor Teillac during my sabbatical at the University of Paris-Orsay. The Laboratory of Neutron Physics was under the direction of Professor I. M. Frank. Its main facility is a pulsed fast neutron reactor. The Laboratory of Theoretical Physics has several theoreticians well-known in the West. The Laboratory of Nuclear Reactions under the direction of Professor G. N. Flerov was of the most interest to me. This laboratory operated two heavy-ion cyclotrons. This group had been very active over the years in the synthesis and investigation of the properties of transuranic elements. I had previously met Flerov several times at international scientific meetings. In addition, his able assistant, Yu. Ts. Ogenession spent several months at the University of Paris-Orsay during the time I was there. We stayed at Dubna, the jewel of Soviet Laboratories, until the afternoon of February 10 and then returned to Moscow.

On the morning of Friday, February 11, we flew from Moscow to Karkov in the Ukraine. In Karkov, we visited the Physical and Technical Institute of the Academy of Sciences. Although this institute was badly damaged during World War II, it had been completely repaired. It has a very large overall staff of approximately 5,000 employees. This institute was some thirty-five years old at the time of our visit and had research programs in the areas of High Energy Physics, Nuclear Physics, Plasma Physics, and Solid-State and Low-Temperature Physics. One item in the Karkov Laboratory of particular interest to me was a heavy-ion linear accelerator, a replica of the machines operating in Berkeley and at Yale. However, the Karkov scientists were not successful in accelerating heavy-ion beams. It was astonishing to see this excellent facility, occupying a large hall, but serving only as a relic gathering dust.

During the late afternoon of Saturday, February 12, we were scheduled to fly to Kiev. Severe storms of wind and snow resulted in the cancellation of all flights. Our KGB shaman came to our rescue. Within minutes, he flashed an ID to railroad officials and was able to acquire ten berths on the fully booked overnight train from Karkov to Kiev! We spent Sunday, February 12, sightseeing in Kiev. On Sunday afternoon there was a wedding in the hotel in Kiev in which we were staying. I took several pictures of the wedding party with an instant camera and presented the developed pictures to the participants. They had never seen such instant photographs and thought it was some kind of special trick. They were really appreciative and thanked me profusely.

Monday, February 14, we spent at the Institute of Physics of the Ukrainian Academy of Sciences in Kiev. This institute, which is smaller than the one in Karkov, was under the direction of Madame A. F. Prikhotko. The research here was mainly in the fields of nuclear physics and solid-state physics although they had some effort in the areas of plasma physics and electronics. Professor O. F. Nemets, a name I was familiar with, had a research program on a small cyclotron. Professor A. S. Davydov had recently come to Kiev from Moscow University and was one of the outstanding theoreticians at the institute.

The large amounts of snowfall continued to be a problem for us. On Tuesday, February 15, we were snowbound in Kiev and were unable to leave by plane. Our itinerary included a trip to Novosibirsk in Siberia. This is largest city in Siberia and third largest city in the USSR. In 1943, the Academy of Sciences opened up the Siberian branch in Novosibirsk. Many research institutes are located in Novosibirsk itself, and many more are clustered in Academgorodok, a city founded in the 1950s by the Academy and located some 30 kilometers south of Novosibirsk. At its height, Academgorodok was the home of 65,000 scientists and their families, and was a privileged area to live in, with well-stocked stores and dachas for the scientific elite. It was a big disappointment, not to be able to visit this complex of scientific institutes in its glory days. In recent years, Academgorodok has fallen on hard times due to slashes in government funding for science.

At this point, because of the severe weather and the cancellation of all air flights, our delegation decided to split into two groups. One group would visit the Khlopin Radium Institute in Leningrad and the other group would visit Obninsk, a large science city of many research institutes, created to explore the production of nuclear energy. I was in the group to visit the Radium Institute in Leningrad. However, many years later in 1987 I spent a week in Obninsk. On Tuesday, February 15, our group took the evening train from Kiev to Moscow and on Wednesday, February 16, the train from Moscow to Leningrad. On Thursday, February 17, our group visited the Khlopin Radium Institute. This was that first institute in the USSR to study radioactivity and has the distinction of constructing the first cyclotron in Europe.

The Khlopin Radium Institute had almost 1,000 employees at the time of my visit, and some 350 scientists. Spontaneous fission of uranium-238 was discovered in 1940 in this institute. One of the large groups studying high-energy nuclear reactions was directed by Professor N. A. Perfilov, a USSR scientist whom I had met some years earlier at a Nuclear Chemistry Gordon Conference in New Hampshire. Udo Schröder visited this institute some tens of years after I did, and one of the scientists remembered that they had had an earlier visitor from the

University of Rochester and proceeded to give him some pictures of me taken during my visit of 1966.

On the evening of Thursday, February 17, our group boarded the overnight train from Leningrad to Moscow. On Friday, February 18, both groups visited the I. V. Kurchatov Atomic Energy Institute in Moscow. This was the largest atomic energy laboratory in the Soviet Union and employed at the time of our visit about 6,000 people including 1,500 scientists. The research programs at the Kurchatov Institute were extensive and highly diversified.

After two weeks of travel within the USSR, rigid schedules of meeting after meeting, intense discussion sessions, sometimes with language difficulties, lengthy banquets and evening entertainment, I was exhausted. The last evening in Moscow our group dined by ourselves in a fine restaurant with tables tightly packed. As an illustration of the vodka problem at that time, I saw something I'd never seen before in a restaurant. An occasional person would pass out, rest his head on the shoulder of a person at an adjacent table, recover after a period of time, and resume his intake of vodka.

Our hosts often pointed out to us the shops in the USSR that were exclusively limited to customers with hard currency. On visits to these shops, I purchased several gift items including nested doll sets (including as many as nine wooden dolls), lacquered wooden boxes and cups, Fabergé eggs, and a balalaika. Including these items enhanced the amount of my luggage for the trip back home.

We left Moscow on Saturday, February 19, for Paris. However, because of severe weather conditions, we were flown first on a smaller plane to Leningrad and then onto Paris. Again we stayed overnight in Paris and then returned to New York on Sunday, February 20. As a prank on the trip home, our delegation concocted an abstract for a talk about our trip, supposedly to be given at a future meeting of the American Physical Society. Our group signed the abstract as "The Kiev Collective" with a footnote giving our names and affiliations. It was all in fun and the abstract never reached the printing press.

Our two-week in-depth look at the state-of-science as practiced in the finest laboratories in the USSR was an unparalleled experience for me. I learned first-hand that Soviet science in general lacked up-to-date equipment and in many ways their science was inferior to that in the United States. Knowing this, I was often annoyed by several administrations, especially the Reagan Administration, in their spreading false propaganda and hype about the USSR, in order to increase the military budget.

On March 30, 1966, soon after returning from the USSR, I received a letter from President Lyndon Johnson informing me that he had approved the recommendation of the Atomic Energy Commission and the General Advisory Committee that I be granted an Ernest Orlando Lawrence Memorial Award in recognition of my achievements in Nuclear Fission Research. This award is reserved for those making a significant scientific contribution prior to their forty-fifth birthday. A follow up letter from the Secretary of the AEC informed me that the award would be presented in Washington, DC, on Wednesday, April 27, 1966. This letter also asked for names of family, friends, and associates whom I wished to receive invitations to the ceremonies. My family accompanied me to Washington to receive the award. The Washington and Chicago Operations Office of the AEC both sent out packages of information of my award to local newspapers and to professional magazines. As a result of this surge of publicity, I received numerous letters of congratulations, some from former colleagues and acquaintances going back into my distant past.

Following my sabbatical in Paris in the fall of 1965, I served for a year on a five-member panel appointed by the National Academy of Sciences-National Research Council to review the status of Nuclear Chemistry. In 1966 our lengthy and comprehensive report (95) was published by the National Academy of Sciences. The content of our report included (I) Introduction, (II) History and Major Accomplishments, (III) Current Trends in Nuclear Chemistry, (IV) Present Organization of Research and Training in Nuclear Chemistry, (V) Future Requirements, and (VI) References.

In Section III, for example, the goals, intellectual content, and status of present-day nuclear chemistry research are outlined. Subsection (1) of this section discusses research aimed at understanding the properties and transformations of nuclei. Subsection (2) turns attention outward from the world of nuclei to nuclear chemical research directed at atomic and molecular problems. Subsection (3) examines the applications of nuclear chemistry to the solution of problems in the macroscopic universe.

Although this report deals principally with pure research, the role of nuclear chemistry in the nation's nuclear energy programs is so basic that it could not be omitted; this role is discussed briefly in subsection (4). Finally, in subsection (5), attention is focused on the interplay between advances in instrumentation and the development of research in nuclear chemistry.

On October 5, 1965, I received a letter from Alexander Zucker, Chairman of the Organizing Committee for the International Conference on Nuclear Physics inviting me to present a paper on Fission at their next meeting to be held on September 12–16, 1966, in Gatlinburg, Tennessee. This meeting was to be the successor to the meetings in Paris (1964) and Padua, Italy (1962). I chose as the subject of my lecture, "The Transition State in Nuclear Fission." My lecture was published (96) in the proceedings of the conference.

Cross-section ratios for isomeric pairs scandium-44 and 44m and cobalt-58 and 58m produced by the potassium-41 (α,n) and manganese-55 (α,n) reactions, respectively, were measured as a function of alpha-particle energy (93). The experimental results were compared to calculations based on different nuclear models in order to study the spin dependence of nuclear level density.

Fission fragment anisotropies were studied for the uranium-235 (d,pf) reaction at several deuteron energies as a function of excitation energy (100). The anisotropy values of less than unity at low excitation energy are interpreted in terms of target deformation and spin.

The last major program initiated while at Argonne was a study of the structure of the fission transition nucleus by measurements of fission fragment angular distributions for fission of even-even actinide targets induced with low-energy monoenergetic neutrons. The fission thresholds of such targets exceed the neutron binding energy, and these reactions offer an opportunity for characterizing the transition-state spectrum, since only a few states in the transition nucleus will be accessible and the properties of the levels in the target nucleus populated by neutron emission are known.

The monoenergetic neutrons were produced from the lithium-7 (p,n) reaction with protons accelerated in the Argonne Van de Graaff. Because of the low intensity of fission fragments near threshold, a highly efficient method of measuring fission fragment angular distributions had to be developed. The main feature of our experimental procedure was the use of a solid-state nuclear track detector to measure the fission fragment angular distributions. It is well known that the radiation-damaged sites caused by fission fragments entering a number of insulating materials can be enlarged by chemical etching until they can be seen with an optical microscope. By choosing a suitable material, the number of fission events can be recorded uniquely within a high background of low-mass particles.

After some experimentation, the detector arrangement chosen can be seen as illustrated in figure 20A, where the detector material is a polycarbonate resin (Makrofol). Monoenergetic neutrons impinge on an isotopically pure even-even actinide target of 0.5 milligram per square centimeter tilted at an angle of 33 degrees to avoid absorption of the fragments in the target. The overall resolution of the neutrons was approximately 15 keV. The detector material was 200 microns thick. It was arranged in the form of a truncated cone at the base, supporting a cylindrical section, which in turn supported a top cone. The top cone was sliced at an angle of 33 degrees so that an elliptical section is at the top near the polar angle of zero degrees (see figure 20).

The reasons for choosing this geometry are (a) it guarantees that all fission fragments from a source at the center of the base of the bottom cone will enter the detector material at angles between 20 and 70 degrees, thus insuring proper track registration; (b) the symmetry of the detector insures rapid reading of the data since all the tracks along a circular ring perpendicular to the axis of symmetry (the beam axis) will correspond to the same angle θ; and (c) it affords the 2π geometry necessary for measuring angular distributions in the low cross-section region near threshold. Both the target and the detector were enclosed in a thin-walled, evacuated aluminum scattering chamber. After irradiation, the detector material was chemically etched (6 normal sodium hydroxide for 40 minutes) and the resulting fragment "tracks" were viewed with an optical microscope. The angular distribution of fission fragments for neutron-induced fission of even-even nuclei is determined by the quantum numbers I, K, M, and π (see figure 10)

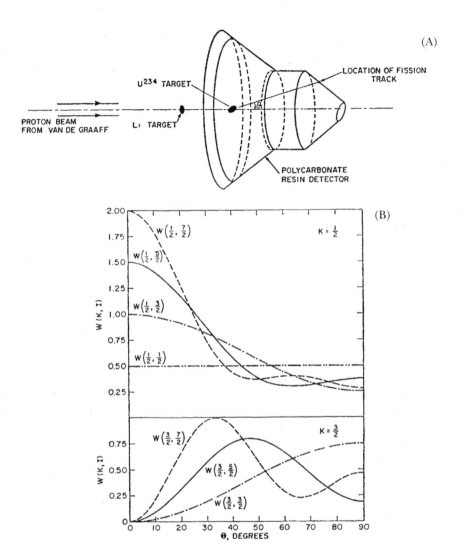

Figure 20. (A) Schematic diagram of the polycarbonate resin detector used for measuring fission fragment angular distributions of fission induced with monoenergetic neutrons (102). (B) Theoretical fission-fragment angular distributions for neutron-induced fission of even-even target nuclei. The different curves are identified by the quantum numbers, K and I, for each transition state (102).

and given by Equation (1) of reference (102). Illustrations of theoretical fission fragment angular distributions for neutron-induced fission of even-even target nuclei are shown in figure 20B. The axis of quantization is along the beam direction and M has values of ± 1/2. The top part of figure 20B is for fission through transition states in a band with K=1/2 and I values of 1/2, 3/2, 5/2, and 7/2. The lower part of figure 20B is for fission through transition states in a band with K=3/2 and I values of 3/2, 5/2, and 7/2. Each curve is normalized such that

$$\int_{-1}^{+1} W(K, I)(\theta)d\cos(\theta) = 1.$$ [21]

This figure demonstrates the strong dependence of the fission fragment angular distributions from neutron-induced fission of the even-even targets on the quantum numbers of the intermediate transition state.

The first (n,f) reaction studied of an even-even actinide target with our new polycarbonate resin detector was uranium-234 (99, 102). To demonstrate the makrofol detector technique, we compared our measurements of fission fragment anisotropies from the uranium-234 (n, f) reaction, at several neutron energies, with previous measurements of Lamphere (see figure 18). Excellent agreement was obtained. Having demonstrated the technique, we proceeded to study the fission fragment angular distributions for the uranium-234 (n,f) reaction at incident neutron energies of 200, 300, 400, 500, 600, 700, 843, 998, and 1184 keV (99, 102). Experimental fission fragment angular distributions at seven of the lowest neutron energies are shown in figure 21 (102). The solid and dashed lines are theoretical fits to the data with five and four accessible deformed-Nilsson states, respectively, in the transition-state nucleus. Sven G. Nilsson first studied the matrix of a single nucleon in a deformed potential well, hence, the notation Nilsson states. Such a transition state is characterized by its K value and parity (e.g., ½+), the energy above the fission barrier (e.g., E_o=450 keV), and the barrier curvature (e.g., $\hbar w = 275$ keV). The notations used to characterize Nilsson states are discussed in a later section entitled Single Particle Nilsson States. The barrier curvature is inversely proportional to the thickness of the barrier. The solid line fits to the data, shown in figure 21, are for the following five accessible Nilsson states (3/2+, 375, 275), (3/2-, 550, 300), (1/2+, 600, 625), (1/2-, 750, 150) and (1/2-, 725, 400). Each independent-particle transition state K supports a rotation band I=K, K+1, K+2 . . .

Following the publicity of the Lawrence Award, invitations from universities escalated. asking me to consider faculty professorships, including even the positions of department chair and college dean. During this time period in my career, I was not interested in pursuing positions with heavy administrative responsibilities. However, I did visit several universities to investigate their facilities for faculty interested in nuclear science. Up to this time, I had chosen to remain at Argonne. The Chemistry Division was ably administered by Dr. W. M.

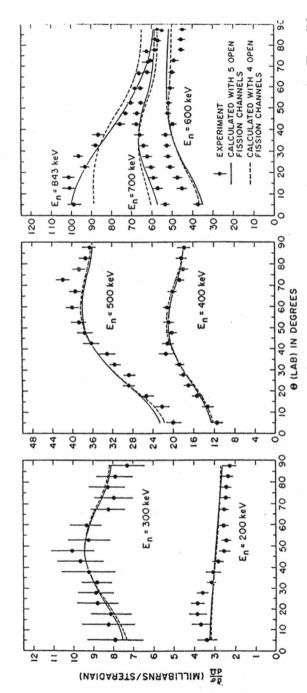

Figure 21. Fission fragment angular distributions for the uranium-234 (n,f) reaction at seven incident neutron energies. The solid points are the experimental data and the curves represent fits with four and five accessible states of the transition nucleus (102). The parameters for the five state fit are given in the text.

Manning, who incidentally had been a PhD student of Professor W. A. Noyes at Brown University. Manning managed the division in an informal university style with group leaders assuming complete responsibility for their own research programs. I fared well in such a system. My group was composed of mostly temporary employees, postdoctoral associates, and graduate students. Manning was very supportive of my group, and I received ample budget for my group's salaries as well as adequate support for new scientific equipment and nuclear detectors.

While I was on sabbatical leave in Paris, Dr. Manning was promoted to Associate Director of Argonne and Dr. M. S. Matheson became the Director of the Chemistry Division. Coincidentally, Matheson was also a PhD student of W. A. Noyes, however, by this time, Noyes was at the University of Rochester. Matheson continued to be very supportive of my research, however, beyond his control, the AEC at this time sharply reduced the funding for the national laboratories.

In order to make the necessary reduction in budget and lay off as few permanent employees as possible, Argonne made the decision to sharply reduce the number of temporary employees. This change in policy would impact directly my mode of research. I had had PhD students doing their thesis research with me from Chicago, Northwestern, Illinois, and Pennsylvania State, and a steady flow of postdoctoral associates from the United States and abroad. Hence, at this juncture in time, I made the decision to move to a university and apply directly to the AEC (or other government agency) for funding of my research group.

Two pending offers, at that time, of the most interest to me were from the University of Rochester and the State University of New York at Stony Brook. Each of these institutions had a nuclear laboratory and each was in the process of installing a new tandem Van de Graaff accelerator. However, Rochester was getting a state-of-the-art Emperor model, capable of accelerating high-resolution beams of helium-ions to energies of over 30 MeV. In addition, Rochester's Nuclear Structure Research Laboratory, which housed the Emperor tandem, possessed an Enge split-pole magnetic spectrograph for high-resolution particle spectroscopy. These special facilities were of particular interest to me in order to embark on a new research program on the study of single-particle states of deformed actinide nuclei. During the time that I was considering the Rochester offer, I conferred several times with W. A. Noyes, a consultant and frequent visitor to Argonne. Noyes, a distinguished statesman of science, was former chair of the University of Rochester's Chemistry Department from 1930–1955. With the encouragement of Marshall Blann, Harry Gove, and Bob Marshak, I accepted a position of Professor of Chemistry and Physics at the University of Rochester in May 1967, to commence on September 1.

The summer months of 1967 were extremely busy, finishing as many research projects at Argonne as possible and preparing for my future research program at Rochester. During a vacation period, I prepared a lengthy research proposal of my planned future research at Rochester and submitted it to the AEC for funding starting in September, 1967. As part of my startup funds, the

University of Rochester agreed to fund the construction of a beam line at the tandem and a scattering chamber. Hence, I supplied a detailed set of engineering drawings of the scattering chamber for construction in the machine shop of the Nuclear Structure Research Laboratory. In addition, I was able, in this short time interval, to recruit three postdoctoral associates to start on September 1, 1967.

On a trip to Rochester, we located a suitable building lot, selected a house plan, and a building contractor. This allowed time for the preliminary building details to be completed in the summer such that house construction could begin on our arrival in September. We were successful also in finding a furnished rental house large enough to store our furniture during the building process. The rental house was conveniently located near the schools and our building lot. The sale of our home in Western Springs was accomplished without a real estate agent.

During my final months at Argonne, I focused on completing the data-taking phase of several experiments, deferring all computations and preparation of manuscripts until my move. In mid-August, the Chemistry Division held a farewell luncheon for me at a nearby restaurant. This terminated for me a very productive eighteen-year association with Argonne. I was sad to leave. Argonne had been good to me, and the Argonne years had been good for my career. During the initial decades following World War II, the national laboratories played a prominent role in elevating the status of science in the United States to the leading position in the world. In the future years, I continued my relationship with Argonne, both as a consultant and as a member of their various advisory panels.

VII. University of Rochester Years (1967–1973)

We arrived in Rochester in late August of 1967 and settled into a rented house in Penfield near the high school. Linda, by this time, was a sophomore in college, Jann and Robert were in their senior and sophomore years, respectively, in high school, and Joel was in eighth grade. Dolly was soon spending a considerable amount of her time supervising the construction of our new home. Everyone in the family adjusted to their new lifestyle quickly. Jann had an Illinois driver's license and made a hit with new friends of her age who were still not eligible for a New York license. Rob made the football team in Penfield as a sophomore, whereas it would not have been possible in his previous high school of 4,000 students, some ten times the size of Penfield. Joel immediately found eighth-grade friends living on the same street as our building lot, one of whom had a tennis court and swimming pool.

My first year at Rochester was a transitional period during which I was still spending a considerable amount of time preparing manuscripts of my previous Argonne experiments. However, with the availability at Rochester of high-resolution beams of energetic helium ions and an Enge split-pole magnetic spectrograph, I also embarked immediately on a systematic study of single-particle and collective states of actinide nuclei by reaction spectroscopy. As an illustration of the short time it took us to initiate and have preliminary publishable results from this completely new research program at Rochester, we submitted our first manuscript on the results from the uranium-238 (helium-3, α) uranium-237 reaction in April of 1968. Up to this time for the benefit of typing without superscripts, I've used the notation (element, dash, mass number) to specify the isotope of the element under discussion, for example, uranium-238. In this section I will use the present notation in the literature, namely ^{238}U. The above reaction is then written as ^{238}U (^3He, α)^{237}U. The symbol preceding the parenthesis is the target, and the symbol after the parenthesis is the heavy-mass reaction product. The first symbol inside the parenthesis is the accelerated projectile interacting with the target. The second symbol inside the bracket is the light-mass reaction product, in this case an alpha particle (a $^4He^{++}$ ion).

A. Reaction Spectroscopy of Actinide Elements

The actinide nuclei represent a rich and fertile region of the periodic table to test theoretical models of the properties of highly deformed nuclei. One of the reasons that I chose to go to Rochester was that the university's Nuclear Structure Research Laboratory had the experimental facilities necessary to study the excited states of these nuclei by nuclear reaction spectroscopy with helium ions.

The angular momentum of odd-A deformed nuclei is due both to the rotational angular momentum and to the angular momentum of the odd nucleon. The projection of the nucleon's angular momentum on the symmetry axis adds vectorially to the rotational angular momentum R to give the total angular momentum I. The projection of the total angular momentum I on the nuclear symmetry axis is given by K. For the prolate actinide nuclei, which are axially symmetric, the rotational angular momentum R is perpendicular to the symmetry axis. When the rotational angular momentum is zero, I and K are equal.

The motion of a single nucleon in a deformed well has been studied theoretically by the Copenhagen school, especially by Sven G. Nilsson. An odd nucleon in a given shell model state can have several different projections of its angular momentum on a symmetry axis. For example, a $g_{9/2}$ single nucleon can have K values of ½, 3/2, 5/2, 7/2, and 9/2. For a prolate odd-A nucleus with axial symmetry with an even-even core (e.g., thorium-231 with one neutron and an even-even core of 90 protons and 140 neutrons) in its lowest energy state, the various single-particle Nilsson levels are represented by a half-integral numeral K, a sign indicating the parity, and three numerals in a bracket [N, n_z, Λ], which are approximate quantum numbers for an infinitely distorted, axially symmetric harmonic-oscillator potential. The first quantity in the bracket N is the principal quantum number (the number of nodes in the wave function). The second quantity n_z is the number of axial nodal planes in the axial direction. The third quantity Λ is the component of the orbital angular momentum I along the symmetry axis.

The first excited Nilsson level of thorium-231, for example, is characterized by 3/2 + [631]. The origin of this Nilsson level is from a $2g_{9/2}$ orbit which for a deformed nucleus splits into the following Nilsson levels: ½ + [640], 3/2[631], 5/2+ [622], 7/2 + [613], and 9/2+ [604], in order of increasing energy. Each of the Nilsson levels can form the ground state of a rotation-band. For example, the above 3/2+ [631] level has a rotational band, 5/2+ [631], 7/2+ [631], 9/2+ [631]. The rotation of the nucleus is around an axis perpendicular to the symmetry axis, and the K value for each rotational state in the band remains unchanged from the ground state of the rotational band.

Alternate notations for the identification of Nilsson levels are also used frequently in the published literature. For example, other notations for the 3/2+ [631] level are 3/2+ [631↑] and [631↑]. The latter notation [631↑] is the most compact. The parity is not given explicitly since this information is contained in the principal quantum number, which in this illustration is even, hence, positive. The upward arrow indicates that the nucleon spin of one-half adds to the orbital angular momentum, which in the illustration is one, giving a total angular momentum of 3/2. In this compact notation, the rotational states built on the ground state Nilsson level [631↑] are designated as 5/2+ [631↑], 7/2+ [631↑], 9/2 + [631↑], and so on.

In the following discussion, excitation of Nilsson levels by transfer reactions on even-even targets is considered. Neutron-hole states are excited predominately with the (^{3}He, α) and (d,t) reactions. Each of these reactions picks up a bound neutron from the sea of neutrons in the potential well of the target. Since the Q value of the (^{3}He, α) reaction on heavy elements is large and the helium-3 and alpha particles are strongly absorbed at the nuclear surface, high angular momentum transfers are favored in this reaction. Many of the neutron single particle states of actinide nuclei arise from $1j_{15/2}$, $1i_{11/2}$, and $2g_{9/2}$ shell-model states, and the wave functions of these states contain a large percentage of high angular momentum. Therefore, one expects that the strongest peaks in the spectrum for the (^{3}He, α) reaction correspond to high-spin levels of the final nucleus. Such high spins, on the other hand, are rarely excited strongly either by radioactive decay or by deuteron-induced transfer reactions. Proton-hole states are excited predominately by the (t,α) reaction in which a proton is removed from the target nucleus. Proton-particle states, however, are excited predominately by the (^{3}He, d) and the (α,t) reactions where in each reaction a proton is added to the target nucleus.

We have also employed the (d,d′) and (p,t) reactions to study the collective excitations of deformed nuclei. The collective states of actinide nuclei that are most strongly excited by inelastic scattering of deuterons are the members of the ground-state rotational band and the low-lying octupole bands. In even-even nuclei the octupole vibration is split by the nuclear deformation into four rotational bands of spin projection K=0, 1, 2, and 3.

1. Single-Particle Nilsson Levels

With even-even deformed targets, the (^{3}He, α) and (d,t) reactions were employed to study neutron-hole Nilsson levels. In the actinide region of the periodic table, measurements were made with targets of thorium-232 (110, 119, 167), uranium-236 (111), uranium-238 (101, 116, 119), and plutonium-242 (120). The nuclides uranium-236 and plutonium-242, enriched to 99.6 and 99.8 percent, respectively, were obtained from the Isotopes Division of the Oak Ridge National Laboratory.

The target material was deposited onto a 20 microgram per square centimeter carbon film. The target was in a rectangular shape with dimensions of 1 millimeter width and 5 millimeters length. The target thickness ranged from 20 to 50 micrograms per square centimeter. Beams of 30-MeV helium-3 ions and 17-MeV deuterons were obtained for the experiments from the Emperor tandem Van de Graaff accelerator of the University of Rochester. Alpha-particle and triton spectra from the (^{3}He, α) and (d,t) reactions, respectively, were analyzed with an Enge split-pole magnetic spectrograph and detected with nuclear emulsion plates, usually 100 micrometer KO Ilford Nuclear Research plates. Spectra were recorded at 20, 25, 35, 60, and 90 degrees in order to obtain angular distributions of the more intense alpha-particle groups.

A scintillation counter mounted at 45 degrees to the beam direction monitored the helium-3 ions and the deuterons elastically scattered from the target. The output of this detector served for intensity normalization of the individual spectra and for the determination of absolute reaction cross sections. The emulsion plates were scanned under calibrated microscopes in strips of one-quarter millimeter width. An automatic computer code was developed to evaluate spectra that included the search for lines, least-squares fitting of a preset line shape, and the calculations of differential cross sections and excitation energies with kinematic corrections.

The triton and alpha-particle spectra, measured at a laboratory angle of 60 degrees, are shown in figure 22 for the ^{242}Pu(d, t)^{241}Pu and ^{242}Pu(^{3}He, α)^{241}Pu reactions, respectively (120). The numbers labeling the individual lines in figure 22 correspond to those given in Table I of reference (120), where the excitations energies and Nilsson assignments of the states in plutonium-241 are listed. In making Nilsson assignments, use is made of the transferred orbital angular momentum obtained from the (^{3}He, α) angular distributions and from the cross section ratio R=dσ(^{3}He, α)/dσ(d,t). The ratio R varies by a factor of a thousand for a change in orbital angular momentum l from zero to seven, and, hence, provides a sensitive measure of the angular momentum transfer. Experimental angular distributions for the 15/2-[743↑] and 15/2-[752↑] states are compared with the theoretical calculations for a l=7 angular momentum transfer in figure 23A and a similar comparison is made for the 9/2+[622↑] state for an angular momentum transfer of l=4. In the Nilsson model, the [743↑] and the [752↑] single-neutron levels originate from the 1j$_{15/2}$ shell-model state and are predicted to be strongly coupled. Coriolis mixing between components of the 1j$_{15/2}$ state has been found in other neighboring actinide nuclei. A partial level scheme of plutonium-241 as obtained by combining our results with previously published data is shown in figure 23B. Levels indicated by heavy lines are obtained in the present experiment. Assigned states seen in previous works but not excited in our experiment are drawn as thinner lines. As seen in figure 23, rotational bands associated with the following Nilsson configurations are identified in plutonium-241: 5/2+ [622], 7/2+ [624], 1/2+ [631], 7/2- [743], 3/2+ [631], 1/2- [501], and 5/2- [752].

Similar studies were made with the (^{3}He, α) and (d,t) reactions on even-even targets of uranium-238, uranium-236, uranium-234, and thorium-232 to study the predominately populated Nilsson hole states in uranium-237 (101, 116, 119), uranium-235 (111), uranium-233 (176), and thorium-231 (110, 119,

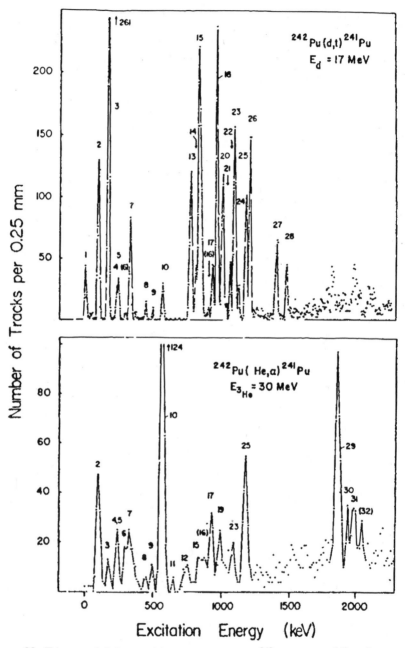

Figure 22. Triton and alpha-particle spectra from the ^{242}Pu(d,t) and ^{242}Pu(^{3}He,α) reactions, respectively, measured at a laboratory angle of 60 degrees. The numbers labeling the individual lines correspond to those given in Table I of reference (120).

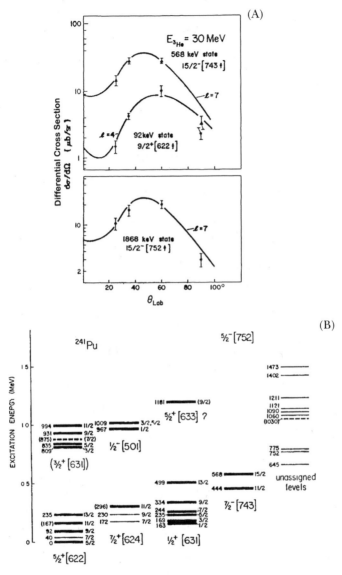

Figure 23. (A) Angular distributions of three alpha-particle groups measured in the ^{242}Pu(^{3}He,α)^{241}Pu reaction. The solid curves are the results of theoretical distorted-wave-Born-approximation calculations using an optical model (120). The ordinate gives the differential cross section in microbarns per steradian. The abscissa gives the laboratory angle in degrees at which the alpha particles are detected (120). (B) Partial level scheme of plutonium-241 as obtained with the present results and previously published data. Assigned states observed in the present experiment are drawn as heavy lines, states seen in previous experiments as thinner lines (120).

167). Angular distributions and cross sections for seven rotational states of three Nilsson configurations from the thorium-232 (d,t) reaction, at a laboratory bombarding energy of 17 MeV, were measured at some twenty laboratory angles and the results compared with different theoretical calculations (167). This study led to the following conclusions: (a) Form factors generated by a deformed Saxon-Woods potential reproduce the observed relative cross sections better than spherical Saxon-Woods well eigenfunctions when used in conjunction with Nilsson coefficients. (b) The shape of the angular distributions is insensitive to the choice of different form factors because the reaction process is localized in the model-independent tail of the radial wave function. (c) Inelastic effects are of minor importance in this reaction. (d) Optical potentials derived from elastic scattering on neighboring spherical nuclei give results when used in (d,t) reaction calculations which are similar to those from potentials derived from an analysis of scattering on deformed nuclei using strong coupling calculations. (e) Absolute (d,t) cross sections calculated with theoretical models are too small.

Proton-particle Nilsson states are predominately populated by the (^{3}He, d) and (α, t) reactions on even-even targets. These reactions were utilized on targets of thorium-232, uranium-234, uranium-236, uranium-238, and plutonium-242 to study Nilsson levels in protoactinium-233 (149), neptunium-235 (179), neptunium-237 (114), neptunium-239 (148), and americium-243 (118). As discussed previously for single-neutron transfer reactions, single-proton transfer reactions also provide a sensitive means of studying heavy deformed nuclei. The distribution of strength among the levels of a rotational band depends on the structure of the intrinsic state, and hence, the measured differential cross sections give direct and detailed information about the nuclear wave functions. Nuclear structure studies based on transfer reactions therefore often compliment similar studies performed by investigating radioactive decay.

As an example of results from the (^{3}He, d) and (α,t) reactions on even-even actinide targets, I'll give some results only for the thorium-232 target (149). The experimental techniques and methods are similar to those mentioned previously. The thorium-232 target was prepared by vacuum evaporation of thorium metal onto 20 microgram per square centimeter carbon backing. To test the target purity, spectra of the (d,d′) reaction were measured at laboratory angles of 90 and 125 degrees. No measurable amounts of heavy-mass impurities were detected. Spectral lines from lighter elements, on the other hand, do not interfere with the (^{3}He, d) and the (α,t) spectra at the angles chosen for these reactions. The (^{3}He, d) experiment was performed with a 28.5-MeV helium-3 beam from the Rochester Emperor tandem, while the (α,t) experiment was done at 30 MeV bombarding energy.

The spectra of deuterons and tritons from the (^{3}He, d) and (α,t) reactions on a thorium-232 target (149) as recorded with an Enge split-pole magnetic spectrograph are shown in figure 24A. The excitation energies and differential cross sections obtained for the various individually numbered states are listed in

Table I of reference (149), along with the proposed Nilsson-state assignments. A partial level scheme of protactinium-233 is shown in figure 24B. The reaction results are interpreted in terms of a distorted-wave-Born approximation analysis. Transferred angular momenta are deduced from (α,t) and (^{3}He, d) cross-section ratios. Theoretical cross sections, calculated in the framework of the Nilsson model with pairing and Coriolis interactions included, are in reasonable agreement with experiment. Rotational bands built on the following Nilsson proton configurations are identified in protoactinium-233: 1/2- [530↑], 3/2+ [651↑], 1/2+ [660↑], 5/2+ [642↑], 5/2- [523↓] and 3/2-[521↑].

A series of experiments were performed also to study predominately Nilsson proton-particle states in the rare-earth region of the periodic table, another domain of highly deformed nuclei. We were able to obtain several nearly isotopically pure even-even targets of gadolinium from the Argonne National Laboratory. These targets of gadolinium-154, -156, -158, and -160 were prepared on the ANL's isotope separator. Utilizing the (^{3}He, d) and (α,t) reactions, these gadolinium targets were bombarded with 26-MeV helium-3 ions and 27-MeV alpha particles to study the Nilsson proton levels in terbium-155, -157, -159, and -161, respectively (130).

Distorted-wave-Born approximation analysis was used to interpret the reaction results, with transferred angular momentum l values obtained from the comparison of the experimental and theoretical values of the cross section ratio, R=dσ(^{3}He,d)/dσ(α,t). The spectroscopic information obtained was compared with the theoretical predictions of the Nilsson model. Corrective terms added included the Coriolis and pairing interactions. Many of the levels excited were identified as members of rotational bands built on Nilsson single-particle configurations. Identified in all four terbium nuclei are the [411↑], [413↓], [532↑], [404↓], [402↑], and [411↓] Nilsson configurations.

The angular momentum, l, discrimination afforded by the comparison of the excitation of a given state in the two different reactions yielded meaningful angular momentum, l, values for many states. The usefulness of this ratio, R, is seen in figure 25A, where the cross-section ratios of states excited in terbium-157 are plotted against the distorted-wave-Born approximation predictions. As can be seen from figure 25, the experimental and theoretical cross section ratios are in excellent agreement for all values of the angular momentum transfer. As an example of the terbium isotopes results, the level scheme of terbium-157 is shown in figure 25B. The thicker lines indicate Nilsson levels populated by the (^{3}He, d) and the (α,t) reactions.

The (t,α) reactions on even-even actinide targets of thorium-230, thorium-232, uranium-234, uranium-236, and uranium-238 were carried out to study the single-proton Nilsson levels in actinium-229, actinium-231, protoactinium-233, protoactinium-235, and protoactinium-237 (164). The (t,α) reaction predominately populates proton-hole states of high spin and, therefore, compliments previous work. A second goal of these experiments is to study the effect of

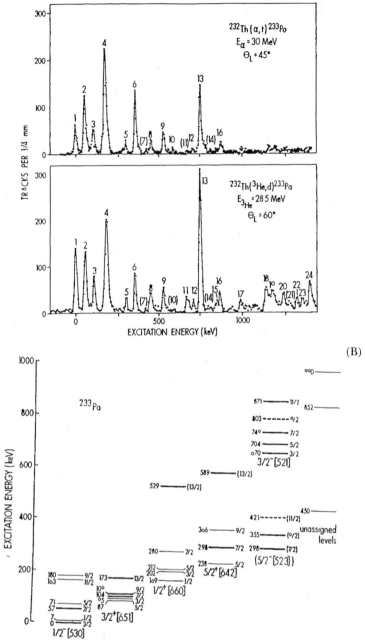

Figure 24. (A) Deuteron and triton spectra from the thorium-232 (³He, d) protoactinium-233 and thorium-232 (α,t) protactinium-233 reactions, respectively. The individual peak numbers correspond to those in Table I of reference (149). (B) Partial level scheme of protoactinium-233. The thicker lines represent levels excited by the (³He, d) and/or (α,t) reactions, while the thinner lines indicate previously identified levels, which are not seen in this experiment. Parentheses indicate uncertain assignments (149).

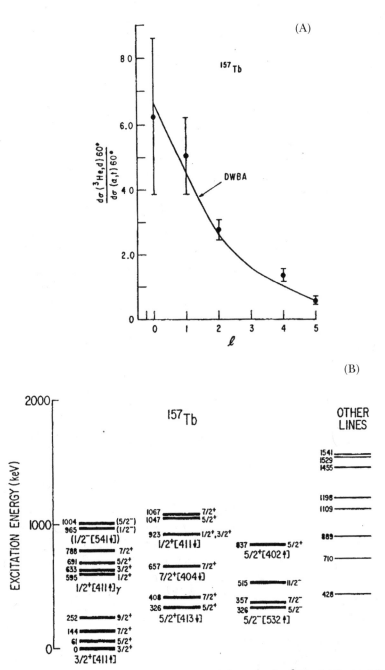

Figure 25. (A) Angular momentum dependence of the ratio $d\sigma$ (^{3}He,α)$/ d\sigma$ (α,t) for the states excited in terbium-157. Excitation energy dependence is removed by adjusting all values to E_{exc} = 1MeV. Points for l=2 and 4 represent averages of two or more ratios (130). (B) Level structure of terbium-157. The thicker lines indicate levels seen in our experiments, while the thinner lines indicate previously identified levels not excited by proton transfer. Parentheses indicate indefinite assignments (130).

increasing neutron number on the proton single particle levels in isotopes of actinium and protactinium. The measurements were performed with a 15-MeV triton beam available from the Los Alamos Scientific Laboratory tandem accelerator. Radioactive triton beams were not accelerated in the Rochester Emperor tandem. The targets have a thickness of approximately 50 micrograms per square centimeter on 20 microgram per square centimeter carbon backing and were prepared by either vacuum evaporation or isotope separator. The scattered alpha particles were recorded by a helical coil detector in the focal plane of a quadrupole-dipole-dipole-dipole spectrograph and typically 1 to 2×10^4 alpha events were detected per run. Spectra for each target were taken at laboratory angles of 50, 60, and 70 degrees. The smallest and largest angles were measured primarily to identify impurities by the mass-dependent kinematic shift. No heavy mass impurities were detected. Absolute cross-section calibrations are based on the experimental intensity of elastic tritons measured with a surface barrier monitor detector of known solid angle located at a laboratory angle of 40 degrees. An energy resolution of 15 to 19 keV full width at half maximum (FWHM) was consistently achieved in all of the runs.

The reaction results are interpreted in terms of a distorted-wave-Born approximation analysis. Theoretical cross sections, calculated in the framework of the Nilsson model with pairing and Coriolis interactions included, are compared with experimental values. The following Nilsson proton configurations are identified in all of the residual nuclei: 1/2+ [400], 3/2+ [402], 3/2+ [651], and 1/2- [530]. The 1/2- [541] and 9/2- [514] configurations are tentatively assigned. Because of the increase in deformation with neutron number, the ground state configuration of the protactinium isotopes changes from 3/2- [530] to 1/2+ [400] as the atomic mass A increases.

2. Collective States

It is well-known that inelastic scattering of medium-energy projectiles predominately excites collective nuclear states. The scattering of complex projectiles, for example, deuterons, helium-3 ions, or alpha particles is especially well suited for studying such states, in as much as compound nuclear processes contribute little to the scattering amplitude. An attractive feature of inelastic scattering processes is the direct relation between the differential cross section observed and reduced transition probabilities. Thus, relative B(E2) and B(E3) values between the ground state and the states excited can be easily determined from such measurements.

The collective states of actinide nuclei that are most strongly excited by inelastic scattering of deuterons are members of the ground-state rotational band and the low-lying octupole bands. Levels of the octupole bands are excited in inelastic scattering through l=3 transfer, and in first approximation only the J^π=3⁻ member of each band should be observed. However, the 1⁻ 5⁻, and 7⁻ members are also seen, primarily as a result of higher-order processes. The cross-section patterns of the octupole bands in (d,d') scattering are quite consistent from one actinide nucleus to another.

We began an extended research program to study the collective states of actinide nuclei by the (d,d') reaction. I'll first describe our deuteron scattering experiments on even-even actinide targets (128, 135, 154). Measurements were made on targets of thorium-230 and -232; uranium-234, -236, and -238; plutonium-240, -242, and -244; and curium-248. Targets were isotopically enriched and prepared by either vacuum evaporation or by isotope separator. The exotic targets were obtained from either the Argonne National Laboratory or the Oak Ridge National Laboratory.

The (d,d') scattering measurements were made with 16-MeV deuterons accelerated in the Rochester tandem. The deuteron spectra were measured at laboratory angles of 90 and 125 degrees in the laboratory. The inelastically scattered deuterons were momentum analyzed in an Enge split-pole magnetic spectrograph and detected with 50 micrometer thick Kodak NTB photographic plates. The energy resolution for the different targets ranged from 8 to 14 keV full width at half maximum (FWHM).

In each of the even-even actinide nuclei, the ground-state band is observed up to and including the 8+ level. Several octupole-vibrational states were identified also in each nucleus. Transition probabilities B(E2) and B(E3) were extracted for selected levels. The B(E3, $0^+ \rightarrow 3^-$) values agree reasonably well with results of a microscopic theory of octupole vibrations that includes Coriolis coupling between negative parity bands. Examples of (d,d') spectra from thorium-230 and plutonium-244 targets are shown in figure 26. One of the striking features of these spectra is the large gap in the spectra of plutonium-244 and curium-248 in the energy region of 500 to 950 keV, where the K=0 band with spins and parity of 1^-, 3^-, and 5^- . . . is located for lighter actinide nuclei such as thorium-230.

Energy spectra of inelastically scattered 16-MeV deuterons were measured at laboratory angles of 90 and 125 degrees using odd-A targets of uranium-233 and -235, neptunium-237, and plutonium-239 (156). These isotopically enriched actinide nuclei were supplied by the Argonne National Laboratory. These experiments extended our investigations of even-even actinide nuclei to odd-A nuclei, which are one nucleon away from even-even nuclei studied previously by (d,d') inelastic scattering.

The (d,d') reaction on these odd-A nuclei is expected to strongly populate members of the ground-state rotational band and the one-phonon states of the core coupled to the target ground state. The ground-state coupling to the one-phonon states give rise to bands with quantum numbers $K_{gs} \pm K_{vib}$. In this region of the actinide nuclei the lowest collective vibration is the $K^\pi=0^-$ octupole state, which coupled to the odd-A ground state, forms a single collective band. In uranium-233, neptunium-237, and plutonium-239 we find this band strongly populated at approximately the energy of the $K^\pi=0^-$ band in the adjacent even-even

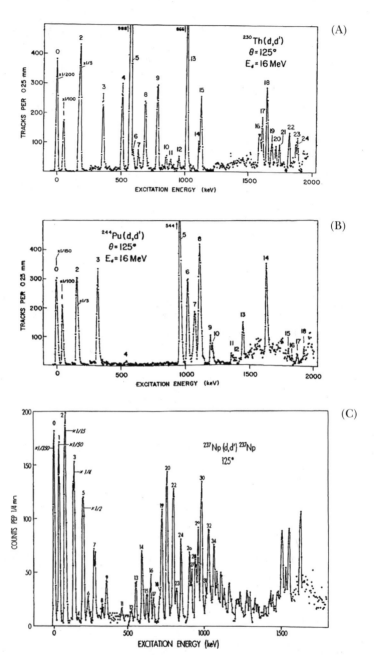

Figure 26. (A) Spectrum of the ^{230}Th (d,d')^{230}Th reaction at a laboratory angle of 125 degrees (154). (B) Spectrum of the ^{244}Pu (d,d')^{244}Pu reaction at a laboratory angle of 125 degrees (154). (C) Spectrum of the ^{237}Np (d,d')^{237}Np reaction at a laboratory angle of 125 degrees (156).

nuclei. And, in fact, the majority of the $K^\pi=0^-$ core collectivity seems to be concentrated in this one band in contrast to being distributed widely among various single-particle states. It is apparent that the ground-state coupling to the higher octupole bands is observed also in the above three nuclei.

The experimental spectrum of the neptunium-237 (d,d′) reaction is shown in figure 26C. The ground-state Nilsson configuration for neptunium-237 of 5/2 + [642] is excited up to and including the spin 15/2 member of this band (numbers 0, 1, 2, 3, 5, and 7 in figure 26C). Some evidence for the 17/2 member of this band is seen in the (d,d′) strength around 300 keV. The collective $K^\pi=0^-$ octupole state is seen in neptunium-237 at 721 keV (peak number 19). The rotational band members 7/2, 9/2, and 11/2 are excited also (peak numbers 20, 22, and 24). The excitation of this band is similar to the J, $K^\pi=3$, 0^- level energies in the adjacent even-even nuclei. There is a second envelope of collective strength centered around an excitation energy of 1 MeV, which corresponds to the ground-state coupling to other collective vibrations.

The angular distributions of elastically and inelastically scattered deuterons from uranium-238 at an energy of 17 MeV are compared with coupled-channel calculations (168). The cross sections at small scattering angles are strongly influenced by nuclear-Coulomb interference effects and allow a simultaneous extraction of nuclear (optical potential) and charge quadruple deformation parameters. Two different deformed Coulomb potentials and the parameters of the optical model are discussed in this article.

The (p,t) reactions on even-even isotopes of gadolinium with mass numbers 152, 154, 156, 158, and 160 were studied with 18-MeV protons (127). The neutron numbers of those nuclei vary from N=88 to 96. Three different situations are covered in this mass region. The ^{152}Gd → ^{150}Gd reaction connects two spherical nuclei, while the ^{156}Gd → ^{154}Gd, ^{158}Gd → ^{156}Gd, and ^{160}Gd → ^{158}Gd reactions connect two deformed nuclei. The reaction ^{154}Gd → ^{152}Gd connects a deformed nucleus to a spherical one.

The angular distribution of the emergent tritons allow an unambiguous identification of l=0 transitions and, in some cases, a classification of l=2 transitions in terms of beta- or gamma-vibrations. Excited collective 0+ states are populated with approximately 15 percent of the ground-state strength. In the ^{154}Gd(p,t)^{152}Gd reaction, a third 0+ state is observed at 1053 keV, which is interpreted as being a deformed 0+ excited state in the spherical ^{152}Gd nucleus.

In an additional study of transitional nuclei, we measured cross sections and angular distributions for a number of levels populated in the ^{188}Os(p,t)^{186}Os, ^{190}Os(p,t)^{188}Os, and ^{192}Os(p,t)^{190}Os reactions (152). Particular emphasis was placed on the identification of $J^\pi=0^+$ levels. Excited levels with spin and parity of 0+ were observed in ^{186}Os at 1061, 1456, 1953, and 1990 keV; in ^{188}Os at 1085, 1477, 1705, and 1823 keV; and in ^{190}Os, at 911, 1543, and 1732 keV. The total percentages of the measured excited l=0 strength relative to the respective ground state strength for the three nuclei are ^{186}Os, 7 percent; ^{188}Os, 11 percent; and ^{190}Os, 11 percent. The relative strengths of the (p,t) ground-state transition for ^{190}Os, ^{188}Os, ^{186}Os are 1:0.80:0.70.

In the above sections on single-particle Nilsson and collective states, we have included a number of papers, which were published as late as 1978, well beyond the cutoff time of 1973. However, it was more convenient to include all of these papers based on high-resolution-particle spectroscopy in this section. A few of these experiments with targets of gadolinium and osmium were also outside the actinide region of the periodic table.

B. Fission Research

1. Fission Fragment Angular Distributions

Fission-fragment angular distributions for the ^{234}U(n,f) reaction were described previously (see figure 21). These measurements were made with a novel 2π geometry polycarbonate resin detector. Similar measurements were made with mono-energetic neutrons for other even-even actinide targets including thorium-232 (104), uranium-236 (113), plutonium-242 (115), and thorium-230 (122). These four reactions excited, respectively, single-particle Nilsson levels in highly deformed odd-A transition-state nuclei of thorium-233 (104), uranium-237 (113), plutonium-243 (115), and thorium-231 (122).

The fission fragment angular distributions and cross sections were measured for the fission of thorium-232 induced by neutrons of energies 1.20 ± 0.02, 1.40 ± 0.02, and 1.60 ± 0.02 MeV (104). The fission cross section and angular distribution both change markedly with neutron energy. Information on the single-particle levels in the transition nucleus thorium-233 is derived by fitting the experimental data with a Hauser-Feshbach type calculation which includes, in addition to the fission channels, the open neutron and gamma-ray decay channels taking into account level-width fluctuations.

Fission-fragments angular distributions were measured for the fission of uranium-236 induced with neutrons of energies 400, 500, 600, 700, 800, 900, and 1100 keV (113). At the lower energies the experimental angular distributions are peaked in the direction of the neutron beam. However, by the time the neutron energy has reached 700 keV the angular distribution has a peak at an angle which is intermediate between the parallel and perpendicular directions to the beam. Again statistical model calculations of the fission cross sections and angular distributions are performed as a function of neutron energy. These calculations are of the Hauser-Feshbach type and include all the open decay channels for fission, neutron, and gamma-ray emission. The number and type (K, π) of transition-state fission channels and their energy parameters are varied in the calculation to give the best agreement with the experimental cross section and angular distribution data. Assignments of quantum numbers to transition states in the highly deformed uranium-237 nucleus and the relative heights of the postulated two-humped fission barrier for this nucleus are discussed.

Fission fragment angular distributions were measured for the fission of plutonium-242 induced by neutrons of energies 500, 620, 730, 990, and 1230 keV (115). The fission fragments were detected by two independent methods, the polycarbonate resin detectors and surface barrier solid-state detectors. The experimental data were analyzed in terms of both a single-particle and statistical model of the intermediate transition nucleus. For the case of plutonium-242, it is not possible to rule out either the statistical or single-particle model on the basis of the comparisons of experiment with the two theoretical models. However, some comments about the two different theoretical calculations and comparison with experiments can be made. With as few as two single-particle states one is able to reproduce the forward-peaked experimental angular distributions at 500, 620, and 730 keV. On the other hand, the experimental data, except those at 730 keV, are fitted well with the statistical model also. The smaller values of K_o^2 obtained with the statistical model theory at the lower neutron energies favor very low excitation energy in the transition nucleus and in a sense favor the single-particle Nilsson model where only one or two low K-channels are contributing. Hence, these results indicate that either the outer barrier is not much lower in energy than the inner barrier or the damping is too weak to appreciably mix the K-channels established by the inner barrier.

Fission fragment angular distributions were measured for the fission of thorium-230 induced with neutrons of energies 682, 715, 730, and 740 keV and energy resolution of 5 keV and energies of 750 and 1000 keV and energy resolution of 20 keV (122). The unusual peak in the excitation function of the ^{230}Th(n,f) reaction for neutron energies in the vicinity of 720 keV is interpreted in terms of a vibrational-mode resonance state in a potential well of a two humped-fission barrier. Evidence is presented to show that the resonance has K=1/2. Theoretical fits to the fission cross sections and angular distribution are shown for both parities.

2. Pairing Gap in the Transition Nucleus

The capture of a neutron in plutonium-239 produces an even-even fissioning nucleus. Studies with resonance neutrons have revealed two groups of fission resonances. The fission widths of each group of resonances is reasonably well fitted with a Porter-Thomas distribution, indicating a single fission channel for each spin state. Furthermore, from the measured average fission width, one deduces that the 0+ channel is completely open, whereas the 1+ channel is only partially open. The 0+ level is thought to define the fission threshold for an even-even transition nucleus. The above information suggests that thermal and resonance neutrons have insufficient energy to excite quasiparticle excitations in the transition nucleus plutonium-240.

The present experiments were initiated to look for the two-quasiparticle threshold in the intermediate nucleus. Fission fragment angular distributions were measured (103, 106) for the fission of plutonium-239 induced with neutrons of energies 150, 400, 475, 550, 600, 700, 800, 900, 1000, 1100, 1200, 1350, and 1500 keV (with energy resolution of 25 keV). Measurements of fission fragment angular distributions as a function of energy reveal information on the transition states of a nucleus as it passes over the fission barrier. These transition states consist of collective and particle excitations of the fissioning nucleus at the so-called saddle-point deformation. In the present experiments, we are especially interested in the first few MeV of excitation energy above the fission threshold. In this energy region, the effects of superfluidity due to the nuclear pairing are expected to be most important. A break in K_o^2 is expected to occur at an energy corresponding to the onset of two-quasiparticle excitations. The values of K_o^2, deduced from the experiments, rise sharply from a value of 5–6, for neutron energies less than or equal to 600 keV, to an approximately constant value of 13, for neutron energies in the range of 1.3 to 2.0 MeV. The break in the K_o^2 curve occurs approximately 2 MeV above the fission threshold and is interpreted as the beginning of two-quasiparticle excitations of the highly deformed transition nucleus plutonium-240. This result suggests that the pairing gap is increased for large nuclear deformations.

Similar measurements of the pairing gap in a transition-state nucleus in a region of the periodic table where the saddle-point nucleus has its largest deformation (see figure 19) are described (109). The compound nuclei polonium-210 and -211 were obtained at excitation energies ranging from 3 to 20 MeV above the fission barriers. The experimental data were fitted by a theoretical expression relating the angular distribution of fission fragments to the distribution of the total angular momentum and of the angular momentum projection K on the symmetry axis at the saddle point. By a least squares fitting procedure, the variance K_o^2 of the K distribution was obtained. In the case of the even-even nucleus polonium-210, the K_o^2 value approaches zero at an energy well above the barrier, while in the case of the odd-A nucleus polonium-211, K_o^2 remains rather large; this indicates the presence of the pairing gap in the polonium-210 and of the residual quasiparticle in polonium-211. Furthermore, the approximately constant difference in K_o^2 between the even-even and odd-A nucleus at corresponding energies is consistent with the expected contribution of a single quasiparticle to K_o^2. From the analysis of the data, the values of the pairing gap at the saddle of polonium-210 is estimated to be larger than in the ground state.

In a later publication (141), the above estimates of the pairing gap at the fission saddle point were reevaluated due to changes in the heights of the second barrier in actinide nuclei. Re-analysis of the polonium-210 data taking into account shell effects at the ground state, but not at the saddle point, has led also to a higher fission barrier. In summary, we concluded in this paper that because of uncertainties in the fission-barrier heights, the above data do not provide an unambiguous answer to the question of whether the pairing strength depends on the nuclear surface area.

3. Other Fission Topics

The ^{209}Bi + ^{7}Li reaction was employed to determine experimentally Γ_f / Γ_n for the initial compound nucleus ^{216}Rn (140). Cross sections for the (^{7}Li, f) reaction were measured at bombarding energies of 24 and 34 MeV by detecting coincident fission fragments. All the other components of the total reaction cross were measured also. Corrections for the effects of second-chance fission were made by lithium-6 bombardments at the appropriate energies. The results give a direct measurement of Γ_f/Γ_n since both fission and neutron evaporation cross sections were determined. The experimental values of Γ_f/Γ_n for radon-216 are analyzed with an angular-momentum dependent statistical model and the effect of incorporating an enhancement of the level density at the saddle point due to coupling to collective rotations was investigated. Fission barriers extracted with a variety of assumptions about the level densities were compared with theoretical predictions.

On November 17, 1970, I gave an invited lecture at the American Nuclear Society meeting in Washington, DC, on the subject Near Barrier Fission Induced with Photons, which was published in the proceedings (125). In this lecture, theoretical expressions are developed for near-barrier fission of even-even targets induced with dipole and quadrupole photons. Although the formulas are developed for a single humped fission barrier and valid when the second barrier has the higher energy, the same equations are applicable for a two-humped barrier by introducing a more complicated expression for the fission transmission coefficients. Dipole fission of an odd mass target is discussed also. Some comparisons are made between theory and experiment. Results are shown of our unpublished data for the ^{236}U(γ,f) reaction, where the anisotropy, W (90 degrees)/W (0 degree), is nearly 10 for the lowest energy with no indication of a quadruple contribution. This is in sharp contrast to the Russian results for uranium-238. Some time later an updated lecture on the subject of threshold photofission was published (134) in the Proceedings of the International Conference on Photonuclear Reactions and Applications.

While on sabbatical leave at the University of Paris-Orsay, we worked on a review article on Nuclear Fission, later published (106) as part of the *Treatise on Nuclear Chemistry* Vol. II (1968), 1–184. During this time period we were also invited to prepare an article on fission for Science magazine (117). We chose the topic "Nuclear Fission Revisited." In this article, I discussed the experimental evidence for a double-humped fission barrier and some of its characteristic consequences, as well as the theoretical basis for the barrier structure.

4. Textbook on Nuclear Fission

While at Argonne in the 1960s, I had entertained the thought of writing a reference book on the subject of nuclear fission, and had made preliminary outlines of the content of such a volume. Encouragement came from some publishers who contacted me stating their interest in such a project. By this time I had published many research papers and written several lengthy chapters (74, 87, 105) in reference books on various aspects of nuclear fission. After my move to the University of Rochester and the establishment of my research program there, I decided that the time had come to move forward on the long-delayed project to write a book on nuclear fission. Recognizing, however, the time commitment of such a project, I decided to seek a co-author. Immediately, Robert Vandenbosch, a professor at the University of Washington, came to mind. He was a former colleague at Argonne, and a co-author of many papers on nuclear fission. He graciously accepted and the writing began.

We wrote *Nuclear Fission* at a level to introduce students to the exciting field of the physics and chemistry of fission. At the same time, we attempted to cover recent developments in nuclear fission at a sufficient depth to make the volume valuable also to research scientists. The theoretical framework for understanding fission is developed in a systematic way and discussed in terms of the most recent experimental observations. A good idea of the book's content (138) can be obtained from the headings of the fourteen chapters, namely, I. Introduction; II. The Fission Barrier; III. Spontaneous Fission; IV. Fission Widths from Neutron Resonance Studies; V. Properties of Low-Lying Levels of Transition Nuclei Determined from Reaction Studies; VI. Fission-Fragment Angular Distributions at Moderate to High Excitation Energies; VII. Competition between Fission and Neutron Emission; VIII. Fission Barrier Heights; IX. Motion from Saddle to Scission: Theories of Mass and Energy Distributions; X. Kinetic Energy Release in Fission; XI. Distribution of Mass and Charge in Fission; XII. Prompt Neutrons from Fission; XIII. Gamma Rays from Primary Fission Fragments; and XIV. Ternary Fission.

This project occupied many evenings and weekends as well as vacation days. I can remember vividly working on revisions and proofs after ski hours at Vail and Aspen, Colorado. *Nuclear Fission,* published by Academic Press, was available to the public in August, 1973.

C. Nuclear Reactions

Excitation functions of the reactions $^{31}P(p,\alpha_0)$ and $^{31}P(p,\alpha_1)$ were measured with high resolution (< 5 keV) at 7 angles in 10 keV steps for proton bombarding energies of 8.37 to 9.00 and 10.00 to 11.77 MeV (107). Auto-correlation functions (see equations [14]–[16]) were calculated from the excitation functions.

From these auto-correlation functions, average widths of the compound states of ^{32}S were determined to be 38±5, 47±4, and 45±5 keV for excitation energies of 18.1, 19.8, and 20.7 MeV, respectively. These widths are compared with statistical model calculations in which level densities of the Fermi-gas type are used. Good agreement between experiment and theory is obtained for level density parameters which give a fair description of the known low-energy level densities of the nuclei which enter the calculations.

The following experiments were performed to measure the effects of isospin on statistical nuclear decay (121, 123, 137). Measurements of the cross sections and energy spectra of protons and alpha particles emitted from the reactions, $^{62}Ni+p$ and $^{59}Co+\alpha$, were made to determine the extent to which isobaric spin is a good quantum number in the intermediate composite system ^{63}Cu at an excitation energy of 18.9 MeV. The composite system, ^{60}Ni, formed by the reactions $^{59}Co+p$ and $^{56}Fe+\alpha$, was investigated also in a similar study at an excitation energy of 22.3 MeV. The experimental ratio, $\sigma\ (\alpha,\alpha')\ \sigma\ (p,p')\ /\ \sigma\ (\alpha,p)\ \sigma\ (p,\alpha)$, was found to be 2.1 for the ^{63}Cu system and 1.6 for the ^{60}Ni system (121). Both values of this ratio exceed considerably the theoretical value of approximately unity predicted by the Bohr independence hypothesis and the statistical model of nuclear decay including angular momentum. The deviations in the above experimental ratios from the compound nucleus predictions are explained by an isobaric spin selection rule which causes the $T_>$ composite system to decay preferentially by proton emission.

In a similar experiment to the one described above, the composite systems ^{64}Zn and ^{66}Zn were investigated, and the measured cross section ratios were 1.3 ± 0.1 and 2.2 ± 0.2, respectively (123). Again the deviation from unity, predicted by the Bohr independence hypothesis, is ascribed to the above isospin selection rule which causes the $T_>$ states to decay preferentially by proton emission.

Statistical fluctuation analyses of the $^{31}P(p, p')$ and $^{31}P(p,\alpha)$ excitation functions to select states give average coherence widths Γ of 30.0 ± 1.4 and 38.7 ± 2.7 keV, respectively, for the composite nucleus ^{32}S at an excitation energy of 17.8 MeV (137). The most probable explanation for the experimental level width Γ from the (p,α) reaction channels being different from that of the (p,p') reaction channels is the partial or complete conservation of isospin in the intermediate composite nucleus. If isospin is a good quantum number, the (p,α) reactions proceed through only the $T_<$ isospin states in the composite nucleus. Hence the level width extracted from these excitation functions is the width of the $T_<$ isospin states. On the other hand, the level width extracted from the (p,p') excitation functions is a weighted average of the widths of the $T_<$ and $T_>$ isospin states. From the experimental values of the average level widths, we determine the level widths of the $T_<$ =0 and $T_>$ =1 isospin states in the composite nucleus ^{32}S to be 38.7 ± 2.7 and 26.2 ± 3.5 keV, respectively. In the case of the composite nucleus, ^{55}Mn, the determined level widths of $T_<$ =5/2 and $T_>$ =7/2 isospin states are 10.8 ± 0.5 and 16.1 ± 1.5 keV, respectively.

In compound nucleus reactions, the relative cross sections for (p,n) and (p,p′) reactions are determined by the available phase space in each reaction channel. As one goes to the more neutron-rich isotopes of a given element, the Q-value for the (p,n) reaction becomes increasingly more positive (i.e., the neutron binding energy in the compound nucleus goes down relative to the proton binding energy) and neutron emission is favored. For sufficiently neutron-rich targets of medium to high Z elements, virtually all the compound nucleus cross section goes into neutron emission, and any observed protons are thus predicted to come predominately from the non-compound nucleus processes. For these reasons, we studied the energy spectra of protons emitted in (p,p′) reactions on targets of ^{118}Sn, ^{120}Sn, and ^{124}Sn as a function of emission angle at incident energies of 14.0 and 17.8 MeV (142). These data confirm the importance of isospin selection rules in such reactions.

D. Nuclear Level Densities

Thick target excitation functions were measured for the reactions ^{59}Co$(p,\alpha_0)^{56}$Fe, ^{59}Co$(p,\alpha_1)^{56}$Fe, ^{59}Co$(p,\alpha_2)^{56}$Fe, ^{55}Mn$(p,\alpha_0)^{52}$Cr, ^{55}Mn$(p,\alpha_1)^{52}$Cr, and ^{62}Ni$(p,\alpha_0)^{59}$Co at two or more angles for proton bombarding energies 6 to 13.5 MeV. The ^{56}Fe$(\alpha,p_0)^{59}$Co reaction was studied at alpha-particle bombarding energies of 12 to 18.5 MeV (112). These measurements were used to determine the parameters in three different level-density formulas. In addition, the experimental values of cross sections to isolated levels were used in conjunction with available values of the level width Γ from cross-section fluctuation measurements to determine level densities of compound nuclei at about 20-MeV excitation energy. The absolute values and energy dependence of the nuclear level density in the excitation energy range 0 to 20 MeV, investigated for several nuclei agree with a back-shifted Fermi gas model.

In August, 1971, an International Conference on Statistical Properties of Nuclei was held at the State University of New York at Albany. I was invited to give a lecture on Experimental and Theoretical Level Densities, later published in the conference proceedings (124). This talk discussed mainly the experimental aspects of the nuclear level density, emphasizing the various methods and techniques employed in the determination of nuclear level densities.

The nuclear state density of a non-interacting fermion system has been calculated directly from a set of single-particle Nilsson levels both as a function of energy and deformation by two numerical methods (126). The first method is a combinatorial calculation and is based on the repetitive use of recursion relations to expand the grand partition function. The second method is based also on the grand partition function, however, in this method one uses the saddlepoint approximation to deduce the relevant thermodynamic quantities. Both methods eliminate the unjustifiable assumption of the conventional theory of level densities that the single-particle level scheme can be represented by a continuous distribution. An approximate relationship between the Strutinsky shell correction and the effective zero of excitation energy extrapolated from the nuclear entropy in a region of high temperature has been demonstrated algebraically. A similar relationship is shown to exist numerically for excitation energies in the 40 to 60 MeV region. The entropy intercept, however, depends on the temperature and is not equivalent to the Strutinsky shell correction for a non-uniform set of single particle levels. In the calculations reported, we emphasize the importance of shell structure on the energy-dependent level density. The results of the calculations show that with a realistic set of single-particle orbitals, one can account qualitatively and in some cases quantitatively for the parameters found necessary to fit experimental data. Calculations reproduce qualitatively the observed values of a_f/a_n found in measurements of the ratio of the fission to neutron emission width and predict also the important role of the liquid-drop fission saddle-point deformation in the interpretation of medium-energy fission fragment angular distributions.

Level densities of ^{56}Fe, ^{59}Co, ^{60}Ni, ^{62}Ni, ^{63}Cu, and ^{65}Cu are determined from analyses of several charged-particle spectra populating the same residual nucleus (129). Each of these experimental spectra is analyzed with an exact theory giving the angular and energy dependant differential cross sections for compound nucleus reactions including explicitly the angular momentum. From these analyses we have determined the Fermi-gas level density parameters: ^{56}Fe, a=5.7 MeV^{-1}, Δ=0.7 MeV; ^{59}Co, a=6.2 MeV^{-1}, Δ=–0.8 MeV; ^{60}Ni, a=6.4 MeV^{-1}, Δ=1.3 MeV; ^{62}Ni, a=6.4 MeV^{-1}, Δ=0.5 MeV; ^{63}Cu, a=6.8 MeV^{-1}, Δ=–0.5 MeV; and ^{65}Cu, a=6.6 MeV^{-1}, Δ=–0.5 MeV.

The analyses of particle-evaporation spectra with the traditionally used approximate theory that neglects angular momentum leads to ambiguous values of the level-density parameter a (131). This ambiguity can be removed by using the exact theory, which explicitly includes angular momentum.

The spin dependences of the nuclear level densities of isotopes of iron, nickel, and copper were studied by measuring angular distributions of particles from (α,α'), (α,p), and (p,α) reactions (132). The experimental values of the spin cutoff parameter σ were compared with theory. When the pairing interaction is included, excellent agreement is obtained between the microscopic theory and experiment.

In 1972, a review article on "Nuclear Level Densities" was published in the Annual Review of Nuclear Science (133). This article contained both (a) a theoretical outline of the methods and models employed in the evaluations of level densities and (b) a listing and description of the experimental sources of information on level densities.

The experimental nuclear level densities and spin cutoff factors for ^{55}Mn, ^{56}Fe, ^{59}Co, ^{60}Ni, ^{62}Ni, ^{63}Cu, and ^{65}Cu are compared with a microscopic theory that includes the nuclear pairing interaction (139). Examples of such comparisons are shown in figure 27, for ^{56}Fe, where part (A) of this figure compares the experimental and theoretical level densities and part (B) the spin cutoff factors. The level densities in figure 27 come from three different types of experimental data. At the lowest energies, the level density comes from counting levels from high-resolution experiments where the individual levels are resolved. At intermediate energies, the level density is derived from measuring charged-particle spectra. At the highest energies, the level density is derived from the measurements of the Ericson fluctuation widths in conjunction with cross sections to isolated levels (112). One sees from figure 27 that at the highest energies, there are over a million levels per MeV.

E. Beryllium-7 Decay under Pressure

In contrast to nuclei that decay by beta- and alpha-particle emission, radioactive decay constants of nuclei that decay by electron capture and internal conversion may be altered by varying the electron density in the vicinity of the nucleus. Hence, such nuclei have been placed in different chemical and physical environments in order to observe these effects.

In our experiments (136), we used diamond anvil presses of a new design to compress samples of beryllium-7 oxide to pressures of 120, 210, and 270 kilobars (one kilobar is equivalent to a pressure of 14,504 pounds per square inch). The press is a simple piston and screw device constructed from stainless steel. The body of the press is basically cylindrical, 2.5 centimeters in diameter and 5 centimeters long. Radiation emitted by the sample emerges through an opening that has a solid angle of approximately 2 steradians. The anvils are 25 milligram (1/8 carat) brilliant-cut single crystal diamonds that have their cutlet faces enlarged to 0.3 and 0.4 millimeter.

Samples of beryllium oxide were prepared from carrier-free isotope in 0.5 normal hydrogen chloride. The pressure samples were made by first compressing a layer of sodium chloride between the diamond faces. Then the piston with the smaller anvil face was removed from the press and a piece of the beryllium-7 oxide in the shape of a platelet about 0.1 millimeter in diameter was transferred to the center of the large anvil. The piston was then reinserted into the press, and the press was loaded to its final pressure.

The purpose of including a layer of sodium chloride between the anvils of the press is twofold. First, it serves as a pressure-transmitting medium and second, it serves as an internal pressure standard. Because of the design of the press, it is possible to use standard X-ray diffraction techniques to determine the lattice parameter of sodium chloride. This value, in turn, can be used to calculate pressure.

The 477-keV gamma ray was used to monitor the decay of beryllium-7. The decay constant λ for the conversion of beryllium-7 to lithium-7 by electron capture

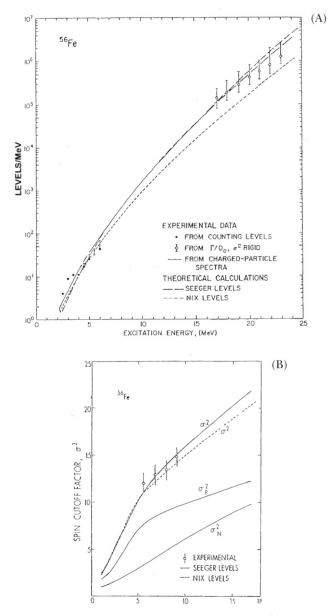

Figure 27. (A) Comparison of the experimental level density of ^{56}Fe with a microscopic theory including the pairing interaction. The theoretical calculations were performed with single particle levels of Seeger et al. and Nix et al. (139). (B) Comparison of the experimental spin cutoff factor σ for ^{56}Fe with a microscopic theory including the nuclear pairing interaction. The individual contributions of the neutrons and protons to σ^2 for the single particle levels of Seeger et al. are shown also (139).

was measured for compressed and uncompressed samples. A least-squares fit of the equation $(\lambda_c-\lambda) / \lambda = K P$ to the experimental data where λ_c and λ are the decay constants of the compressed sample and uncompressed sample, respectively, and P is the pressure in kilobars, yields a value of $(2.2 \pm 0.1) \times 10^{-5}$ kilobar^{-1} for the constant K. The increase in the decay constant with pressure was also estimated theoretically with the electron wave functions of the free beryllium atom, and qualitative agreement was obtained.

F. Other Activities Including Lectures in Japan

The responsibility or principal role of a university professor is often stated to be: "to advance knowledge" through teaching, research, and public service. I learned quickly after coming to Rochester that the third activity, "public service," was open-ended, and I began to be appointed to numerous boards, committees, and panels. Within one year, I received appointments to more than a dozen such committees and panels. For example, the Editorial Board of Physical Chemistry, Advisory Committee to the Editor of the Review of Modern Physics, Standing Program Committee of the American Physical Society, Advisory Council of Princeton's Chemistry Department, and Planning Committee for LAMPF (Los Alamos Meson Physics Facility). Other extracurricular activities during this early period included being the after-dinner speaker at the meeting of the American Nuclear Society in New York City and lecturing at the Nuclear and Space Symposium. I was also invited to join the X-Club, composed of a few members of the University of Rochester as well as prominent members of the Rochester industrial scientific community. This group met monthly with cocktails and dinner preceding a presentation by one of the members.

During my early years at Rochester, I was involved with several study committees and proposals for the construction of a heavy-ion accelerator in the eastern half of the United States. Theoretical studies predicted that an island of rather stable elements may exist far beyond the known elements in the existing periodic state. The theory predicted a shell closure, for example, at Z=114.

To obtain some experimental information on the size of the shell gap at this proton number and therefore on the stability of possible super heavy elements, we attempted to study the (^{3}He,d) and (^{4}He,t) reactions on a target of curium-248. These reactions produce single-particle proton Nilsson states in berkelium-249. The aim was to identify the energy of the expected single-particle proton state, which should be of primarily $f_{5/2}$ character and should therefore, in the spherical limit, originate above the Z=114 shell closure. Its identification would provide information on the size of the shell gap at Z=114 and, therefore, on the stability of possible super heavy elements. With curium-248 being a highly radioactive and hazardous target, we were not successful in positively identifying the Nilsson states coming from the spherical $f_{5/2}$ orbital.

During 1968 and early 1969, we at Rochester prepared a proposal to the AEC and NSF for establishing a regional high energy, very-heavy-ion accelerator. The proposal had the university's approval and was mailed on July 11, 1969, to the above agencies with an accompanying letter of endorsement from President Allen Wallis. Our proposal requested some 6.25 million dollars for capital equipment with the university providing the building. In my Correspondence Volume 1968–77, I have a copy of the letter from AEC Chairman Glenn T. Seaborg to Allen Wallis, received on July 30, 1969, stating that the AEC now has five such proposals for a heavy-ion facility and that he cannot hold much promise for AEC support.

One of the proposals was to convert the Princeton–Pennsylvania accelerator on the Forrestal Campus of Princeton University to have also a heavy-ion capability. I attended several of their meetings over a period of time. On October 10, 1969, I sent a rather critical evaluation letter about this proposal to the AEC raising some technical questions as well as questions about their proposed operating policy. They proposed to maintain a high-flux light-ion capability along with a heavy-ion capability. Such a dual-use machine would not meet the needs of the heavy-ion research community. In the early summer of 1973, I was appointed to a National Academy of Sciences panel that was asked "to review heavy-ion research programs, considering all relevant existing facilities, and attempt to determine quantitatively the scientific need for improving existing, or establishing new, facilities. The Panel also will evaluate accelerator technology applicable for heavy ion accelerators and study any alternatives which, in the Panel's view, would meet the need in heavy ion research." In the end, none of the several heavy-ion proposals that I was involved in were funded.

On May 20, 1968, I was invited by the Japan Society for the Promotion of Science to give a series of lectures over a thirty-day period under the auspices of its Visiting Professor Project. We eventually worked out an agreeable time for me to visit, March 10 to April 10, 1969. I was able to get away during the spring semester since I had a double teaching assignment during the fall of 1968. I taught the large general course in freshman chemistry in which the registration was some 700 students. The large lecture hall in Hutchison holds only some 500 students, requiring two sections for the main lectures. The class was divided into study sections of approximately twenty students for questions and discussion. This teaching arrangement allowed me to devote full time during the spring semester to research and supervision of my graduate and postdoctoral students.

Dolly and I arrived late in the evening and very tired on March 10 in Tokyo. Our hosts insisted they still take us out for a dinner party they had planned in a fancy Tokyo restaurant with geisha-girl entertainment. Among our hosts meeting us at the airport was Professor K. Otozai from Osaka University who had spent a sabbatical year with me at Argonne. Since academic salaries in Japan are low relative to the corporate scale, wealthy industrialists serve as benefactors for professors and pay for their entertaining expenses. This was the case for my visit also and late in my stay I was escorted to the benefactor's presidential suite located on the top floor of his large manufacturing company. Here he presented me with a check covering my travel expenses.

On our first morning in Japan, they had a fluke snow and ice storm. We were scheduled to travel north to Sendai for a lecture at Tohoku University. Not accustomed to such weather, people in Tokyo were on the streets spraying water from hoses attempting to melt the icy mixture. This made the taxi ride to the train station especially hazardous since the drivers were oblivious to the ice. After several near crashes we arrived at the train station. Train travel in Japan is a pleasure with punctual departures and arrivals.

From Sendai we traveled south to Tokai where I gave a lecture at the Japan Atomic Energy Research Institute. One of the highlights of our trip for Dolly was a weekend trip to the pottery site of Mashiko, a world-famous Japanese potter. We then traveled to the Osaka-Kyoto area. My next lecture was at the Reactor Institute of Kyoto University located in Kumatori. At Osaka University, I gave a lecture series of six hours over a period of several days. One of the unique features of our travels in Japan was the sizable welcoming and send-off crowds during our arrivals and departures from the various train stations. Somehow they were able to have a number of their laboratory staff available for such occasions. We soon learned that this was a Japanese tradition. Frequently we would see company employees lined up at a train station seeing the arrival or departure of their supervisor. Traveling to the southernmost island of Kyushu, I lectured at Kyushu University in Fukuoka. There is an active volcano on this island where one is able to walk to the crater's edge and view the smoldering lava unprotected by a fence. We were told it was a favorite site for Japanese suicide. Near the crater were numerous open structures in which to seek immediate shelter from the falling hot lava in case of an eruption. During our travels we passed through both Nagasaki and Hiroshima, the cities where atomic bombs were dropped.

The next lecture was at Kyoto University. We spent several days in this area visiting the old capital city of Kobe. My final lecture was at Tokyo University. During my stay in Japan, the university students were protesting against their respective university administrators. These clashes were the most violent and severe at Tokyo University. Here large battalions of police with metal body shields could be seen openly fighting with the students. These riots caused my lecture to be moved to a research institute in the suburbs of Tokyo. Our final days in Tokyo were at the climax of the cherry blossoms. In Tokyo we stayed at the New Otani hotel, which was very deluxe and had beautiful gardens. On a couple of our stops we chose Japanese accommodations, in order to experience the local traditions.

On December 12, 1968, I received a letter from the Director of Research for the Robert A. Welch Foundation inviting me to participate in their 1969 Conference as a Round Table Discussion Leader. This conference was to be on the subject of the transuranic elements and was being organized with the assistance of Glenn T. Seaborg, the Chairman of the Atomic Energy Commission (AEC). Among the several other people invited was Professor V. Strutinsky, noted nuclear theorist from the USSR. I was asked by the AEC's Division of International Affairs to supervise his stay in the United States and, in particular, to coordinate his extensive travel schedule both before and after the Welch Conference.

Strutinsky's first stop in the United States was at the University of Rochester where he spent two days (November 2–4) and gave the Physics Colloquium. I had met him previously at several international conferences and again on my visit to the USSR in February, 1966. We invited him to stay at our home and an interesting event occurred during the first evening. After a tiring day of scientific discussions and dinner, we arrived home in the dark and I activated our garage door opener and drove into the garage. The next morning at breakfast he voiced his curiosity about the magic of the door opening. He said he worried throughout the night about the garage door and had thoughts about the KGB.

Before the conference, I arranged to have Strutinsky visit the Lawrence Berkeley Laboratory and the Los Alamos Scientific Laboratory. After the conference, he visited Oak Ridge National Laboratory, Princeton University, and the University of Maryland. Arrangements were made with each institution to pay him a stipend sufficient to cover travel and living expenses. His last two days in the United States were spent in Washington, DC A problem of note arose about the Welch stipend. Strutinsky had learned that Sven Nilsson's stipend was larger than his, and this activated his latent capitalistic instincts. I was given the task of soothing his ego by pointing out the Welch Foundation's generosity in not only covering all of our expenses but, in addition, giving us a stipend. After the conference, I was invited to spend a couple of days at Texas A&M University.

In 1969, Phil Abelson, Editor of *Science,* asked me to write a general 5,000 word article for *Science* (117). Among the letters in 1970 inquiring about my interest in available positions, was one to be Physics Chair at the University of Tennessee where the faculty members have available the extensive facilities of the nearby Oak Ridge National Laboratory. Another from Michigan State University invited me to spend the summer, and a more exotic letter encouraged me to consider Strassman's Chair at the University of Mainz. Strassman, a Nobel Laureate and co-discover of nuclear fission, was retiring. In 1970, I was elected to be Chairman of the Division of Nuclear Chemistry and Technology of the American Chemical Society. This position has responsibilities for two years. The first year one serves as the Program Chairman of the Division, organizing the week-long symposia at the national meetings in the spring and fall. The second year one has the responsibility to be the spokesperson for the division and preside over division meetings and all other divisional transactions.

I spent the month of May 1971 as a guest professor at the University of Frankfurt. During this period I gave three lectures on nuclear fission. Hence, this turned out to be an uninterrupted block of time that I could devote almost exclusively to working on our book manuscript described previously. However, I did take time out to fulfill a request to deliver a seminar at the Max-Planck Institute in Heidelberg. In early 1971, I was notified that I was elected to a committee of three to represent the interests of the outside users of the forthcoming SUPERHILAC. The Berkeley HILAC was revamped to have the capability of accelerating very heavy ions, extending over the entire periodic table, and renamed the SUPERHILAC. It was the only such accelerator in the United States and was designated a national facility. On completion of the SUPERHILAC, our group was one of the first to do experiments with very heavy projectiles in November, 1973 (to be described later). In early 1971, we were granted a National Science Foundation Award of some 10^5 dollars that allowed us to purchase a state-of-the-art multiparameter analyzer and associated equipment, which was especially valuable for collection of multi-signal nuclear data.

University and outside committee assignments continued to arrive. Some committees met frequently, for example, the Space Radiation Effects Laboratory's Users Advisory Committee met monthly. Others like Advisory Committees at the National Accelerator Laboratory in Batavia, Illinois, and the Brookhaven Tandem less frequently, while some, like the 1972 Physics Panel for the Lawrence awards, were one-time events. Some of these committee meetings were more enjoyable than others. I found the Advisory Committee at Princeton to be both educational and an enjoyable experience.

The International Atomic Energy Agency (IAEA), with its headquarters in Vienna, Austria, is composed of a group of more than one hundred countries. The IAEA held its first symposium on the Physics and Chemistry of Fission in Salzburg, Austria, on March 22–26, 1965, and a second symposium on the same subject on July 28–August 1, 1969. I attended both of these conferences and presented papers. At my suggestion, the third International IAEA Symposium was hosted by the University of Rochester on August 13–17, 1973. I made two trips to Vienna during the previous year to meet with IAEA officials in the planning of the conference and selecting the speakers. Donald Parry and his staff at the University of Rochester ably handled all of the administrative work and I took responsibility for the scientific matters. Organizing such a conference with more than 300 participants brings on numerous crises. One problem was the weather. We had an uncharacteristically hot week for Rochester and the conferees were located in non-air-conditioned dormitories. I had invited John Wheeler, world-renowned theoretical physicist and co-author with Niels Bohr of the first theoretical paper on nuclear fission, to be the after-dinner speaker. To protect him from the heat, I invited him to stay at our home. The lectures were held in Hutchison Hall and the coffee breaks were held on the veranda on the second floor of Hutchison. The conference turned out to be an excellent one and we received letters of appreciation from the AEC and the IAEA for hosting it. Fortunately, our book on nuclear fission came out in mid-August, just ahead of the conference, allowing it to come quickly to the attention of our peers attending the conference.

VIII. Second Guggenheim Year (1973–1974)

Relieved of the heavy responsibilities associated with the Rochester Fission Conference and the departure of all participants by August 20, I was able to concentrate during the final days of August on the numerous last-minute tasks at the university and to prepare for our departure to Berkeley. We rented our home for a year to the family of a Kodak executive, returning from duty in South America.

I arranged for this sabbatical to be spent in several different locations. The fall semester of 1973, September through December, was spent in Berkeley. The months of January and February, 1974, were spent at the Technical University of Munich, located in Garching, a suburb of Munich, where the tandem accelerator, jointly operated by the Technical University and the University of Munich is located. The four months of March through June were spent at the Niels Bohr Institute, University of Copenhagen, and the final two months of July and August were spent at the University of Munich in the city.

Before leaving Rochester, we had driven Jann, a graduate student, and Rob, a senior, to Ann Arbor where they attended the University of Michigan. On our drive to Berkeley, we dropped off Joel in Boulder, where he was a sophomore at the University of Colorado. We were unsuccessful in locating an apartment in Berkeley for four months; however, the Lawrence Berkeley Laboratory found us a place in Walnut Creek, some ten miles inland over the hills. Most of the time, I drove to the campus. Often Dolly spent the day in Berkeley also as she registered for fall courses in Danish and German.

During my early weeks in Berkeley, I worked on the preparation of two manuscripts on level densities. The first was titled "Comparison of Neutron Resonance Spacings with Microscopic Theory for Spherical Nuclei" (143). Neutron resonance data are the most extensive source of information on nuclear level densities. In this type of experiment the nuclear energy levels are observed at an energy just exceeding the neutron binding energy, and the number of levels is obtained by counting the resonances in a particular neutron energy interval. In this paper, the nuclear level spacings determined from neutron resonance experiments for nuclei with $20 \leq A \leq 148$ and $181 \leq A \leq 209$ are compared with spacings calculated for spherical nuclei with a microscopic theory that includes the nuclear pairing interaction. In such a theory, the state density is calculated with realistic sets of single particle levels by the grand-partition-function method for a system of interacting fermions. The gross features of the experimental data due to nuclear shells are reproduced with the microscopic theory. In addition, the absolute agreement between experiment and theory is reasonable (67 percent of the 151 cases examined agree to within a factor of two) in view of the uncertainties in the experimental data, the theoretical single particle levels, and the pairing strength.

In a second companion paper, we made the same comparison for nuclei with static deformation (144). In this paper, a level density formula including low-energy rotational levels for nuclei with axially symmetric deformation is tested with neutron resonance data for lanthanide and actinide nuclides. The calculations with the microscopic theory, including nuclear pairing, utilize deformed single particle levels of Nilsson et al. The experimental data for the lanthanide and actinide nuclei are consistent with a theory that includes collective rotational levels. The derivation and applicability of a level density formula that includes collective rotational levels is discussed. The above two papers were the subject of seminars given at the Lawrence Berkeley Laboratory, the University of Washington, Oregon State University, and the Lawrence Livermore Laboratory.

Another manuscript worked on during my stay in Berkeley reported on total reaction cross sections of deformed nuclei. The cross sections for alpha-particle scattering and alpha-particle induced-fission of uranium -233 and -238 were measured at bombarding energies of 15 to 27 MeV (145). For these fissionable systems, the fission cross sections are very nearly equal to the total reaction cross sections. These experimental reaction cross sections are compared with various theories based on spherical and deformed potentials in order to investigate the effect of static target deformation on the reaction cross sections. From such comparisons, no effect of target deformation is established. An interaction barrier (defined by $T_{l=0}=1/2$) of 22.34 MeV is obtained from a spherical optical model fit to the experimental reaction cross-section data of uranium. This value agrees within 2.3 percent with values deduced by a number of other methods.

One of my principal aims at Berkeley, before the semester was over, was to have a successful experiment with very heavy projectiles from the SUPERHILAC. As I've discussed previously, none of our many efforts to promote a heavy-ion facility in the eastern part of the United States had been successful. During my stay in Berkeley, the SUPERHILAC was just coming on line and several scheduled experiments during the fall, including ours, were terminated for lack of a useful beam. To my surprise and satisfaction, however, our group had the first successful run on the SUPERHILAC with a krypton beam during the three-day period before Thanksgiving, 1973. My colleagues from the University of Rochester and Argonne came to Berkeley to participate in this experiment, as did Vic

Viola on sabbatical from the University of Maryland. The excellent timing of our successful run allowed me to carry the data to Europe for study and analysis. I had a particularly fruitful collaboration with scientists in Copenhagen about the interpretation of the data to be discussed later.

With the successful run on the SUPERHILAC with krypton-84 projectiles on a bismuth-209 target, my stay in Berkeley was a great success. During the fall semester, Dolly and I were guests at several faculty homes, including Thanksgiving dinner at the home of Helen and Glenn Seaborg. During the Christmas holidays, Jann and Joel came for a visit and we went skiing at the Lake Tahoe and Heavenly Valley ski resorts. Early in January, I by chance saw Professor Jorrit de Boer in the cafeteria at the Lawrence Berkeley Laboratory. He had just arrived from Munich and was to spend a few months in Berkeley. He needed a car and I had one for sale, so we decided on the spot to exchange cars. He bought my car in Berkeley and I would pick up his car on arriving in Munich. This exchange worked out very well and on leaving Munich, I resold the car to de Boer.

On our flight east, we stopped in Chicago to visit friends and, in particular, to see a Michigan–Northwestern wrestling match. Rob was a star on the University of Michigan's team and he won his match easily in a very short time. From Chicago we flew to Zurich, Switzerland, where we boarded a train for Villars, Switzerland. Our traveling luggage was cumbersome as it included skis and boots. In Villars I attended, and was one of the lecturers at, the European Conference on Nuclear Physics. Villars, a small and picturesque village in the Alps, noted for its excellent skiing, was for several years the site of this Winter School. The lectures were held during the mornings and the evenings with the afternoons free for skiing. Indeed, a very civilized schedule for learning. After the meeting in Villars, we traveled by train to Munich. Our apartment in Garching was near the campus of the Technical University as well as the accelerator facilities of both universities. Near the end of the month of January, I gave a talk on level densities of spherical and deformed nuclei to a joint colloquium of both universities.

A paper on our experimental results from lithium-ion bombardments on a bismuth-209 target at the Rochester tandem was prepared for publication during this period in Munich. In this paper we report cross sections for producing various isotopes of radon, astatine, polonium, and bismuth, following bombardment of bismuth-209 by lithium-6 or lithium-7 projectiles in the laboratory energy range of 25 to 34 MeV (146). Because of their proximity to closed shells, most of the nuclei formed in these reactions are short-lived alpha-particle emitters. The lithium ion beams were pulsed with a repetition time of 400 nanoseconds and the alpha-particle decays were observed between beam pulses. In order to eliminate prompt background, only alpha particles were accepted in a semiconductor detector placed at 90 degrees to the beam which occurred in a 100 nanosecond interval beginning 200 nanoseconds after the beam pulses.

In March we moved to Copenhagen. The Niels Bohr Institute located an apartment for us in a city where housing was scarce. After settling in at the Bohr

Institute, I made trips by train to Vienna, Austria, and Marburg, Germany, to give colloquia. The week of March 18–22 was spent in France. The first four days visiting my old colleagues at the University of Paris-Orsay and Friday at Saclay. In Copenhagen, I turned my attention to heavy-ion reactions. The data from our Berkeley experiment in which bismuth-209 was bombarded with krypton-84 projectiles were analyzed and a draft of a manuscript on these experimental results was prepared. In a short communication to *Physical Review Letters* (147), our abstract stated: "Detailed information on a new reaction mechanism is presented for a 600-MeV ^{84}Kr bombardment of ^{209}Bi. This reaction process represents a major fraction of the total reaction cross section and is characterized by strong damping of energy degrees of freedom, but on average, a relatively small amount of mass exchange compared to fission." In this experiment, we use a large position-sensitive detector (PSD) subtending an angle of 26 degrees in the reaction plane, in conjunction with a 2-degree acceptance defining detector. This permitted the measurement of both angles and both energies of fragments in a binary event with a high geometry factor. Four parameters of data were recorded for each event, including (a) the energy deposited in the defining detector; (b) the energy deposited in the PSD; (c) a signal proportional to the position in the PSD; and (d) a signal related to the difference in flight time of the two fragments. The masses and center-of-mass kinetic energies of the reaction products were calculated off-line using the measured laboratory energies and angles of the coincident products.

We labeled this new reaction process "strongly damped collisions." It is characterized by unique mass, energy, and angular distributions. The process is binary resulting in two heavy-mass products with masses centered near those of the projectile and target. The total kinetic energies of the products corresponds to the Coulomb energies for charge centers separated by 15 to 17 fermis (a fermi is 10^{-13} centimeters), a stretched configuration similar to that encountered in fission. An important feature of the angular distribution of the lighter projectile-like fragment is its sharp peaking at an angle slightly smaller than the grazing angle, indicating a relatively fast process.

In Copenhagen, we developed a classical model, with nuclear stretching under the influence of a repulsive conservative force and a radial force, to explain the kinetic energy spectra and angular distributions observed in our ^{209}Bi + ^{84}Kr experiment (150). Our model is motivated by observing the behavior of inelastic collisions between liquid drops. As soon as the drops touch, they start to interact at a distance corresponding approximately to the sum of their radii. As the drops separate, a neck is formed and energy is dissipated until the system snaps at a distance that is considerably larger than the touching distance.

According to the classical picture, we assume that nuclei approach each other on a trajectory determined by the conservative and dissipative forces. For very large initial orbital angular momenta, projectile and target fail to reach a critical distance at which they interact strongly and, therefore, elastic scattering or few-nucleon transfer occurs. However, for initial angular momenta smaller than the limiting value l_{SDC}, projectile and target reach a distance where the motion is highly damped due to strong dissipative forces. At this point a neck develops and the systems continue to rotate while stretching under the influence of a repulsive conservative force and a retarding radial friction. Finally, as the distance between the charge centers reaches the snapping distance, the fragments separate and move apart under the influence of the repulsive conservative force. In this paper (150) we derive reaction times of 0.5, 1 and 2 x 10^{-21} seconds for l values of 240, 200 and 160, respectively, from the ^{209}Bi + ^{84}Kr reaction at a bombarding energy of 600 MeV. These times are comparable to or shorter than those estimated for the descent from saddle to scission in nuclear fission.

On weekends we visited some of the tourist sites in Denmark outside Copenhagen, such as the Fredensborg Castle and the Fredensborg Palace, the summer residences of the Danish Queen. The old town in Aarhus is also an interesting place to spend some time. Legoland is a small village constructed entirely of Lego Blocks. Of special interest to me was our visit to the Bohr Cottage in northern Denmark. Margrethe Bohr, the wife of Niels Bohr entertained us as she still actively maintained the cottage. Outside, on the cottage property, I saw the small building that Niels Bohr used for writing while at the cottage. I mentioned in a previous section that I had met Niels Bohr in 1958 (he died in 1962). Aage Bohr, son of Niels, and his wife, Marrietta, were also present. Niels and Aage are one of the rare parent-child pairs to win Nobel Prizes.

While in Copenhagen, we traveled by bus with a group from the Bohr Institute to attend the Nobel Symposium in Ronneby, Sweden, organized by Sven Nilsson of Lund University.

In early June, I traveled by train to Berlin to give a colloquium on our heavy-ion research at the Hahn-Meitner Institute. Later in the month, I made a trip to EURATOM in Geel, Belgium, to give a seminar on nuclear level densities. Returning to Munich in early July, I gave colloquia on our heavy-ion-reaction research at research institutes and universities in Jülich, Darmstadt, Heidelberg, and Munich. A large part of my remaining time in Munich was spent on the preparation of a series of lectures entitled Selected Aspects of Very Heavy Ion Reactions, presented at the VII Summer School of Nuclear Physics (September 2–13, 1974) held in Mikoljki, Poland. These lectures were later published (151). I had received the invitation to give these lectures in December, 1973. The school is located on the shore of a lake in the vicinity of a large forest in the Masurian lake district in northern Poland. We drove from Munich to Poland and made visits to the University of Krakow, an interesting old city, and to the University of Warsaw before the school opened. After my lectures, we returned to Munich, delivered our car back to Professor de Boer, and departed quickly by air for Rochester.

IX. University of Rochester Years (1974–1983)

During this time period, a large portion of our research time was spent on the study of heavy-ion reactions. We did, however, initiate an additional research program investigating actinide muonic atoms at the Los Alamos Meson Physics Facility.

A. Reaction Spectroscopy

Several papers (148, 149, 164, 167, 176, and 179) on the study of single-particle Nilsson states by high-resolution reaction spectroscopy were published during this time period. These papers were mentioned earlier in Subsection VII.A.1, where this subject was discussed. Similarly, papers (152, 154, 156, and 168) on the study of collective states by high-resolution reaction spectroscopy published during this time period were discussed previously in Subsection VII.A.2.

It was during this time period that our highly successful research program on the investigation of the excited states of heavy nuclei, especially actinide nuclei, by high-resolution reaction spectroscopy was terminated. A summary of the nuclei studied by this technique is given in Table 1. Seven different direct reactions were employed in these studies, including the (p, t), (d, d'), (d, t), (t,α), (^{3}He, d), (^{3}He,α), and (α, t) reactions. The numbers in the table refer to my publication list reproduced in the Appendix.

B. Heavy-Ion Reactions

During this time period, intense excitement was generated by nuclear theorists who had predicted an island of enhanced nuclear stability for elements with atomic numbers much larger than those of known elements. These calculations

Table 1. Nuclei studied by high-resolution particle spectroscopy.
The numbers refer to my publication list reproduced in the Appendix.

Nuclide	(p, t)	(d, d')	(d, t)	(t, α)	(^{3}He, d)	(^{3}He, α)	(α, t)
Gadolinium-150	127						
Gadolinium-152	127						
Gadolinium-154	127						
Gadolinium-156	127						
Gadolinium-158	127						
Terbium-155					130		130
Terbium-157					130		130
Terbium-159					130		130
Terbium-161					130		130
Osmium-186	152						
Osmium-188	152						
Osmium-190	152						
Actinium-229				164			
Actinium-231				164			
Thorium-230		154					
Thorium-231			119			110	
Thorium-232		128					
Protoactinium-233				164	149		149
Protoactinium-235				164			
Protoactinium-237				164			
Uranium-233		156	176			176	
Uranium-234		135					
Uranium-235		156				111	
Uranium-236		135					
Uranium-237			119			101, 116	
Uranium-238		128, 168					
Neptunium-235					179		179
Neptunium-237		156			114		114
Neptunium-239					148		148
Plutonium-239		156					
Plutonium-240		154					
Plutonium-241			120			120	
Plutonium-242		128					
Plutonium-244		154					
Americium-243					118		118
Curium-248		154					
Berkelium-249					Annual Report		

leading to the predicted properties of superheavy nuclei were based in part on experimental information about shape-dependent nuclear shells which accounted for the then-recently discovered two-humped fission barrier.

The search for superheavy elements was one of the initial motivations for my interest in the various types of mechanisms involved in the interaction between two large complex nuclei. It turned out, however, that these reaction mechanisms are of great fundamental interest in themselves. In this section, emphasis will be placed on a new reaction process first recognized in 1973–74, which we called strongly damped collisions. The process is known by various names, however, including deep-inelastic transfer, quasi-fission, and relaxed processes.

In order to place this new damped-reaction process in perspective with other known reaction types, figure 28A shows an overall classification scheme for nuclear collisions (158, 172, 228.) It is based on the classical concepts of a well-defined impact parameter, a small spreading of the wave packet about an average classical system trajectory, and a shape of the intermediate dinuclear complex that develops with increasing classical interaction time. The neglect of a potential interference or correlation of partial waves with very different angular momentum l implies a localization of different reaction types in l space, an assumption underlying the scheme of figure 28. Here, different angular momenta are represented by the corresponding distances of closest approach, r, whose relation to an interaction radius R_{Int} determines the reaction characteristics. Experimental systematics suggest that nuclear interactions become important at center separation distances smaller than R_{Int}, corresponding to nuclear surface separations smaller than about three fermis (1 fermi=10^{-13} centimeter).

The classification scheme in figure 28A illustrates development of the reaction flow, the dinuclear shapes, and the resulting reaction characteristics from top to bottom according to decreasing angular momentum and distance of closest approach, but increasing interpenetration and increasing classical interaction time available to the various relaxation and equilibration processes occurring in the reaction. Distant collisions with $r > R_{Int}$ do not lead to nuclear reactions. Here, Coulomb excitation and elastic scattering are the only contributions to the total cross section. For somewhat smaller angular momenta associated with r values of the order of the interaction radius, $r \approx R_{Int}$, inelastic scattering and exchange of a few nucleons are induced in peripheral, grazing-like collisions. Consequently, there is already some loss of kinetic energy of relative motion, that is, the reaction Q value is slightly negative. For smaller l values with a distance of closest approach significantly smaller than R_{Int}, the interacting nuclei come into solid contact. A window opens between the reaction partners allowing a considerable exchange of mass and damping of the relative kinetic energy. Nevertheless, the constituents of the intermediate dinuclear complex

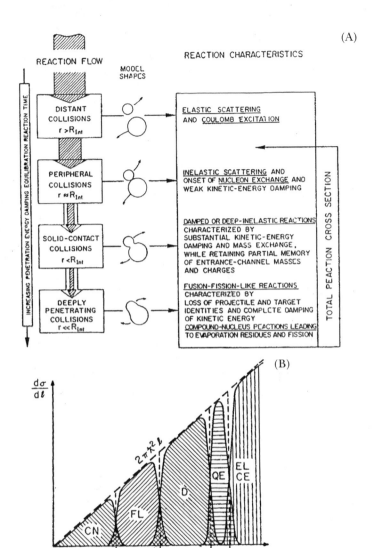

Figure 28. (A) Classification scheme of collisions between complex nuclei based on a classical impact parameter model. The quantity r denotes the distance of closest approach associated with a given impact parameter. Its relation to the interaction radius R_{Int} determines the characteristics of the reaction described on the right. The model shapes apply to the distance of closest approach. The variation in the widths of the hatched vertical arrows illustrates the dominantly peripheral character of collisions between very heavy nuclei. Lighter systems yield deep interpenetrations more readily, as is represented by smaller variations in the widths of the open arrows (228). (B) Schematic illustration of the l dependence of the partial cross section for compound-nucleus (CN), fusion-like (FL), damped (D), quasielastic (QE), Coulomb-excitation (CE), and elastic (EL) processes. The long-dashed line represents the geometrical partial cross section $d\sigma/dl = 2\pi\lambda^2 l$. Vertical dashed lines indicate the extensions of the various l windows in a sharp cutoff model with the characteristic l values noted at the abscissa. Hatched areas represent the diffuse l windows assumed in a smooth cutoff model (228).

retain their separate identities to a large extent, meaning, they carry a memory of the entrance-channel conditions. This memory is lost completely in deeply penetrating fusion-fission-like collisions where the window between the reaction partners becomes so large that a single nucleus with a continuous interior is produced, rather than a dinuclear system. In addition, the kinetic energy of relative motion is completely damped out. A subset of these collision types corresponding to the longest interaction time may eventually produce a compound nucleus characterized by a statistical equilibrium of all degrees of freedom, which may subsequently decay by fission or the evaporation of light particles.

The reaction flow indicated by vertical arrows in figure 28A is distributed between various collision types, accounting for the total reaction cross section in rather different ways, depending on the mass of the projectile-target combination and the bombarding energy. For a relatively light system (open arrows) at not too high bombarding energies, a considerable fraction of the reaction cross section is associated with fusion-like collisions and compound-nucleus production. As the bombarding energy is raised, this contribution loses significance, due to an increased centrifugal repulsion. For heavy systems (hatched arrows), the strong Coulomb repulsion prevents deeper interpenetrations such that compound-nucleus formation becomes negligible. For these latter systems, solid-contact collisions result from a broad band of l waves providing reaction products with a continuous range in the extent to which the various degrees of freedom are relaxed.

As schematically illustrated in figure 28A, increasing penetration and energy damping for heavy-ion reactions leads to different reaction processes. The sum of the individual cross sections for these processes make up the total experimental reaction cross section that is usually analyzed in terms of an effective maximum initial angular momentum l_{max} leading to a nuclear reaction. By employing a sharp cutoff model, the total reaction cross section can be written as

$$\sigma_R = \pi \lambdabar^2 (l_{max} + 1)^2. \qquad [22]$$

Classically, the total reaction measures the sum over the geometrical cross sections $\sigma(l) = (2l + 1)\pi\lambdabar^2$ corresponding to each of the contributing partial waves l up to l_{max}. As is illustrated in figure 28B, various reaction types such as compound nucleus (CN), fissionlike (FL), damped (D), and quasielastic (QE) processes are assumed to correspond to successive angular momentum windows exhausting the total reaction. In a smooth cutoff model, different reaction types may compete for l values in the transition region between adjacent l windows.

1. Elastic Scattering

In the determination of the cross sections for the various reaction mechanisms in heavy-ion reactions it is important to have an estimate of the total reaction cross section σ_R in order to determine the relative probability of each mode of interaction and to insure that all possible partial reaction cross sections have been found. Estimates of the total reaction cross section can be made from the analysis of elastic scattering angular distributions using an optical model, or from strong absorption models that parameterize the partial wave reflection coefficients. For very heavy-ion reactions, however, where the Coulomb force is of similar magnitude to the nuclear force, an analytic expression for the elastic scattering angular distribution can be derived from a model based on a Fresnel scattering analysis. In addition, the total reaction cross section can be calculated from the "quarter-point" angle technique. In this model the maximum angular momentum l_{max} for a particle taking part in a reaction with the target is given by,

$$l_{max} = \eta \cot(\tfrac{1}{2}\theta_{1/4}) \qquad [23]$$

where η is the Coulomb or Sommerfeld parameter equal to $z_p z_T e^2 [\mu/(2\hbar^2 E_{cm})]^{1/2}$ and the angle $\theta_{1/4}$ is the quarter-point angle obtained from the experimental angular distribution as the angle for which $\sigma_{el}/\sigma_{ruth} = 0.25$. The reaction cross section is then given by Equation [22].

Elastic scattering experiments between heavy nuclei are scarce where the elastic channel is measured with sufficient energy resolution to isolate it from the inelastic channels. High-resolution spectra of inelastically scattered [6,7]Li ions from [238]U and [232]Th showed considerable cross section for exciting the low-lying 2+ and 4+ states of these deformed targets (153). However, comparisons of the elastic and inelastic cross sections with predictions of a Coulomb excitation code indicate that the inelastic scattering of these states is predominately due to Coulomb excitation.

The elastic scattering of [40]Ar from targets of [209]Bi and [238]U was studied at laboratory energies of 286 and 340 MeV (155, 159). Particle detection was achieved with two silicon surface barrier detectors mounted on moveable arms. One detector (referred to as the defining detector) was covered by a rectangular slit, subtending at the target one degree in the reaction plane and ±2.5 degrees out of the plane. The second detector was a position-sensitive detector, which subtended a total of 25 degrees in plane. The position-sensitive detector was covered with ten rectangular apertures, each of which subtended one degree in plane and 3.7 degrees out of plane. By a similar technique, the elastic scattering of [84]Kr from targets of [209]Bi and [238]U was studied at laboratory energies of 600 and 712 MeV (155, 159).

Because the Coulomb parameter η is large for these systems, the elastic scattering angular distributions show a classical Fresnel scattering shape. The above experimental elastic scattering data were fitted with optical and Fresnel models. The total reaction cross section deduced from the Fresnel model by the one-quarter-point technique agrees within a few percent with the result from the optical model. The Fresnel interaction radius and the optical model strong absorption radius are found to be approximately equal and qualitatively reproduced by the sum of the half-density electron scattering radii of the two heavy ions and a constant of 3.2 ± 0.3 fermis.

The elastic scattering of ^{136}Xe from ^{209}Bi was studied at three laboratory energies of 940 (202), 1130 (180, 160), and 1422 (210) MeV. Again the elastic scattering data were analyzed with the quantum-mechanical optical model and a semiquantal model of Fresnel scattering. A generalized Fresnel model was employed with a smooth Fermi-type reflection function leading to a finite transition region in l-space, where the transmission coefficient increases from zero to one. For the elastic scattering of ^{136}Xe from ^{209}Bi at laboratory energies of 940, 1130, and 1422 MeV, the quarter-point angles $\theta_{1/4}$ are 70.1, 54.3, and 38.0 degrees, respectively. The corresponding reaction cross sections are 2060, 2750, and 3810 millibarns.

The angular distributions for elastic scattering of ^{40}Ar, ^{84}Kr, and ^{136}Xe by ^{209}Bi, obtained by the method described above, are shown in figure 29. The solid lines demonstrate the excellent fits to the experimental data obtained with the generalized Fresnel model. On the basis of the good agreement for the Fresnel and optical models, one concludes that the Fresnel model is extremely useful for estimating the total reaction cross section for heavy-ion collisions. The generalized Fresnel model, despite its much greater simplicity, gives information about the interaction between heavy ions equivalent to that given by the optical model.

2. Strongly Damped Collisions

During this time period we spent a sizable fraction of our research effort in studying the characteristic properties of this newly discovered damped reaction process. The main features of this process at a few MeV/nucleon above the interaction barrier may be summarized as follows:

a.) The reactions are basically binary with only two massive reaction fragments in the exit channel. Subsequently, the excited primary fragments decay via the emission of light particles and/or fission and gamma rays.

b.) The angular distributions of the reaction fragments change in a characteristic way with the charge product $Z_p \cdot Z_T$ of the projectile-target combination and the bombarding energy. For some lighter systems, forward-tending orbiting-type components develop. However, very heavy systems exhibit fragment angular distributions that are focused into a relatively narrow angular range centered close to the grazing angle. Examples of such angular distributions will be shown later for the ^{209}Bi+ ^{136}Xe reaction. There are definite correlations between the angular-distribution type and the net exchange of mass and charge between the

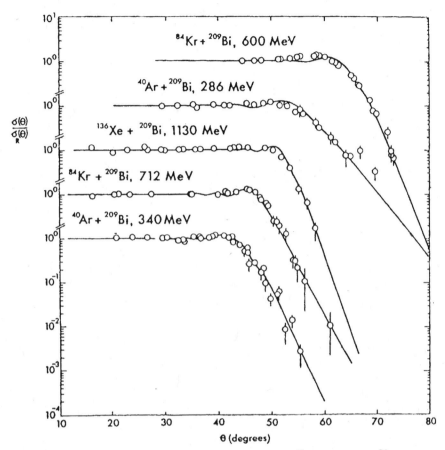

Figure 29. Angular distributions for elastic scattering of ⁴⁰Ar (155, 159), ⁸⁴Kr (155, 159), and ¹³⁶Xe (159, 180) by ²⁰⁹Bi. The solid lines are fits to the experimental data with the generalized Fresnel model (228).

reaction partners, as well as with the amount of kinetic energy dissipated in the reaction. The correlation with the energy loss is the more fundamental one.

c.) Fragment mass or charge distributions are bimodal and centered close to masses and charges of projectile and target nuclei. For narrow bins of kinetic energy loss, mass, and charge distributions are nearly bimodal Gaussians with widths that increase with increasing energy loss. There are also strong correlations between the net amounts of mass and charge transfer between the reaction partners that develop with increasing energy loss.

d.) Large amounts of kinetic energy of radial and orbital motion can be dissipated in a reaction. Final total kinetic energies can be as small as those corresponding to the Coulomb repulsion of two highly deformed fragments. The partition of the dissipated energy between the final reaction fragments is consistent with an intermediate system that is close to a statistical equilibrium at the instant of scission. Final fragment spins can also be large with a significant alignment, which is strongly depended on the total kinetic energy loss.

The damped reaction process has been extensively studied for the ^{209}Bi + ^{136}Xe reaction (157, 158, 160, 180, 202, 210, 213) at three laboratory energies of 940, 1130, and 1422 MeV. Thus, I'll use various experimental results from this reaction to illustrate the above general features of strongly damped collisions. Striking examples for the shape of angular distributions obtained for the above systems are displayed in figure 30A where the experimental differential cross section is plotted versus laboratory angle for the projectile-like fragments. These three laboratory energies are between a factor of 1.3 and 2.0 larger than the Coulomb energy calculated for spherical nuclei touching at the strong-absorption radius. The associated large asymptotic Sommerfeld parameter ranging from $\eta=268$ to 218 suggests that these systems are close to the classical limit. As is the case for most reactions induced by heavy projectiles, an insufficient experimental energy resolution of only 1 to 2 percent prevents a clear separation of quasielastic from damped events. Hence, quasielastic reactions are included in the angular distributions of figure 30. As is obvious from this figure, the cross section is strongly focused into a narrow peak located a few degrees forward of the corresponding grazing or quarter-point angle indicated by an arrow. The concentration of the cross section and its rigid correlation with the quarter-point angle is characteristic of a fast peripheral reaction. It is important to note that the angular distributions shown in figure 30A exhaust the total reaction cross section in each case and, hence, comprise several hundred l waves.

In the case of the ^{209}Bi + ^{136}Xe reaction at the lowest energy of 940 MeV, we were able to measure both the projectile-like and target-like fragments simultaneously. figure 30B shows the laboratory angular distributions for both fragment types, where target-like events are attributed to negative reaction angles. The asymmetry of the distribution of the heavy reaction partners is understood to be due to a kinematical effect. The integrated cross section for these events is somewhat reduced from that of the light fragments because of sequential fission of highly excited Bi-like primary fragments.

The phenomenon of a strong angular focusing of the cross section, so clearly demonstrated for the ^{209}Bi + ^{136}Xe reaction, has an origin that is rather different from that of the grazing peak observed with lighter systems for the quasielastic events. For the latter systems, the cross section peak is due to l waves in the vicinity of the grazing angular momentum (or l_{max}) that experience very little deflection by the nuclear interaction, whereas smaller angular momenta lead to trajectories that are deflected forward by increasing amounts. In contrast, for the heavy system ^{209}Bi + ^{136}Xe, the total cross section is seen to be focused into a narrow angular range. Hence, in this case, the balance between conservative and dissipative forces leads to a classical deflection function that is approximately independent of l, that is, $\theta(l) \approx \theta_o$, for several hundred l waves.

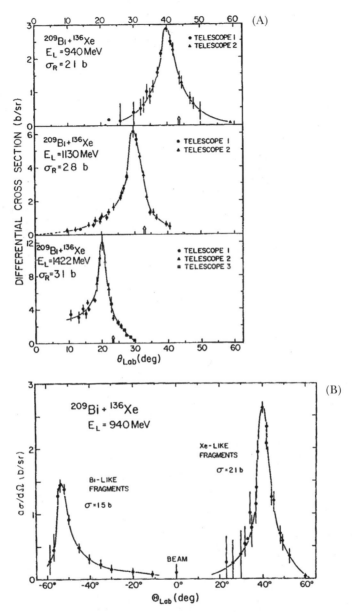

Figure 30. (A) Laboratory angular distributions for the ^{209}Bi + ^{136}Xe reaction at laboratory energies of 940 (202), 1130 (180), and 1422 (210) MeV. The arrows indicate the angle of the quarter point in the laboratory. The center-of-mass energies above the Coulomb barrier for the three cases are 1.75, 3.14, and 5.29 MeV/nucleon. (B) Laboratory angular distributions of light (Xe-like) and heavy (Bi-like) reaction products observed for the ^{209}Bi + ^{136}Xe reaction at a laboratory energy of 940 MeV (202).

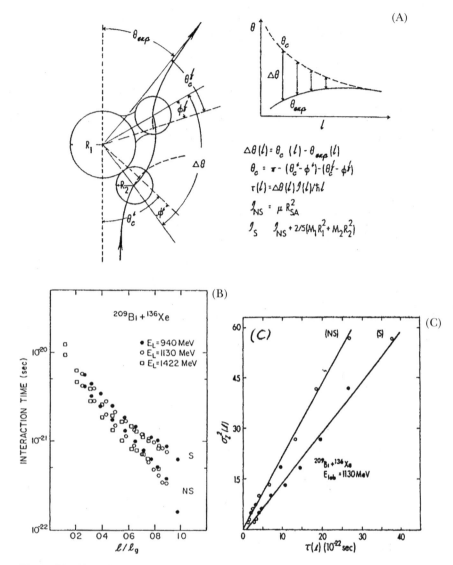

Figure 31. (A) Schematic illustration of the classical model used for the calculation of interaction times (150, 169). (B) The mean interaction time is plotted as a function of the initial relative angular momentum l/l_g for the ^{209}Bi + ^{136}Xe reaction at three laboratory bombarding energies. The final angular momentum is calculated in the limits of the nonsticking (NS) and sticking (S) models (210). (C) The variance σ_z^2 of the experimental fragment-charge distribution plotted as a function of the interaction time τ for the ^{209}Bi + ^{136}Xe reaction. The two sets of results correspond to the assumption of nonsticking (NS) or sticking (S) of the two ions during the interaction (166).

A procedure used to deduce the interaction times of strongly damped collisions (150, 151, 166, 169) is illustrated in figure 31A. Here, the two spherical nuclei are assumed to follow Coulomb trajectories prior to nuclear contact time and after breakup of the intermediate system. The corresponding Coulomb deflection angle $\theta_c(l)$ as well as $\theta\Delta(l)$, the angular difference between the Coulomb deflection angle and the reaction angle $\theta_{exp}(l)$, are illustrated in figure 31A. The angular-momentum-dependent interaction time is calculated with the expression,

$$\tau(l_i) = \Delta\theta(l) \, \mathscr{I}/\hbar l_f \qquad [24]$$

where $\mathscr{I}$ is the moment-of-inertia of the double-nucleus system (see figure 31A).

The l-dependent interaction times for the ^{209}Bi + ^{136}Xe reaction at three laboratory bombarding energies are plotted as a function of l/l_g ($= l_i/l_{max}$) in figure 31B. Here, the nonsticking (NS) and sticking (S) cases refer to conservation of orbital angular momentum ($l_i=l_f$) or to instant reduction of l_i to the sticking value. In either model, the interaction times are seen to increase approximately exponentially with decreasing angular momentum l. We have observed a similar behavior for a number of other strongly damped reactions (166, 169, 172). As can be seen from figure 31B, for a given ratio l/l_g, the interaction time does not depend much on bombarding energy. This interesting result is consistent with the predictions of a transport model to be discussed later. It is understood to be mainly due to the dynamical balance of Coulomb, centrifugal, and frictional forces for heavy systems and bombarding energies well above the interaction barrier.

The observed correlations between the kinetic energy loss and nucleon diffusion to be discussed later suggest that the nucleon-diffusion mechanism operates continuously on the fast time scale of 10^{-22} to 10^{-21} seconds simultaneously with energy dissipation. The variance of the fragment-charge distribution is then also l-dependent and gives the relation $\sigma_z^2(l) = 2D_z(l)\tau(l)$. Experimental values of $\sigma_z^2(l)$ as a function $\tau(l)$ are plotted in figure 31C for the ^{209}Bi + ^{136}Xe reaction for both the nonsticking and sticking models.

There are few features of damped nuclear reactions that have yielded a deeper insight into the microscopic reaction mechanism than those associated with the observed division of the total mass and charge of a system between the fragments merging from a binary heavy-ion collision. Experimental studies of various moments of A, Z, isobaric, and isotopic distributions of damped reaction fragments and their variation with remaining reaction variables such as scattering angle and total kinetic energy loss have initiated the development of theoretical models involving methods that are new to nuclear reaction theory.

Two seminal papers (157, 161) published in 1976 as *Physical Review Letters* were very instrumental in advancing our understanding of damped reactions. In the first paper (157), it was found that the centroids of the charge (Z) distributions for the xenon-like fragments, from the ^{209}Bi + ^{136}Xe reaction at a laboratory energy of 1130 MeV, stay close to the charge of xenon as the energy loss increases. At the same time, we observed a rather dramatic broadening of the charge distribution as the energy loss increases. These important findings are illustrated in figure 32A where the double-differential cross section, $d^2\sigma/dEdZ$, is plotted against the atomic number Z for the ^{209}Bi + ^{136}Xe reaction at a laboratory energy of 940 MeV (202). The Z distributions are plotted for a succession of 26-MeV-wide bins in total kinetic energy, with centroids as indicated at each distribution. These are angle-integrated data. It is rather striking how well the experimental spectra are described by the Gaussian fit curves drawn through the data points. The rather stationary centroids and dramatic broadening of the charge distributions with energy loss, as shown in figure 32A, are suggestive of a classical diffusion process in which nucleons are randomly exchanged between projectile and target nuclei, leading to increasingly wider distributions as the interaction time proceeds. The full width at half maximum (FWHM) of the fragment Z distributions shown in figure 32A has been measured to be independent of reaction angle (202), once a final kinetic energy or energy loss has been chosen (see figure 32B).

In the second seminal paper (161) published in 1976, we reported a correlation between the kinetic energy loss and nucleon diffusion for Kr- and Xe-induced reactions on heavy targets. Although this correlation suggests that these two phenomena occur on the same general time scale, the rate of kinetic energy loss decreases with interaction time. Evidence was presented to show that the initial energy dissipation is consistent with a frictional force that is proportional to the relative velocity.

The correlation between the charge distribution of the projectile-like fragment, σ_Z^2, and the total kinetic energy loss, E_o-E, in damped reactions is shown in figure 33A (161). The quantities E_o and E are the initial and final asymptotic total kinetic energies. The variance increases smoothly with increasing amounts of total kinetic energy loss. In addition, the slope of the energy dissipation as a function of the variance σ_Z^2 is largest for small values of the variance and decreases as the variance increases. Utilizing the reaction times from Equation [24] and the initial slope of the total kinetic energy loss versus σ_Z^2 plot shown in figure 33A, as well as the relationship of σ_Z^2 to reaction time shown in figure 31C, one derives an initial kinetic energy dissipation rate, -dT/dt, of approximately 2.5×10^{23} MeV/second for a heavy-ion collision. This is equivalent to 4×10^{10} watts or 1/15 of the peak power of the US power grid!

Although the four reactions shown in figure 33A have similar correlations between the energy loss and variance in the charge distributions, the dynamic influence of the bombarding energy is clearly visible for large variances and/or energy losses as shown in figure 33B for the ^{209}Bi + ^{136}Xe reaction at the highest energy (211).

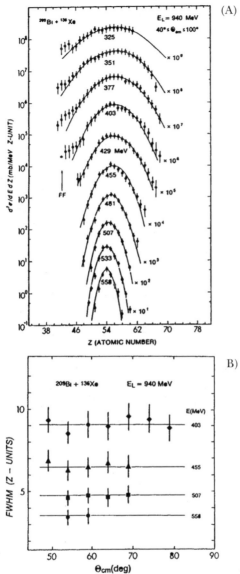

Figure 32. (A) Z distributions of fragments from the reaction ^{209}Bi + ^{136}Xe at a laboratory energy of 940 MeV are plotted as a function of final total kinetic energy indicated at the curves. Energy bins are 26 MeV wide. Solid curves represent Gaussian fits to the data (circles). The distribution for 558 MeV corresponds to elastically scattered xenon ions and illustrates the experimental Z resolution. The arrow (FF) indicates contamination of the data by events from sequential fission of target-like fragments (202). (B) The FWHM of Z distributions $d^3\sigma/d\Omega dEdz$ for the indicated final kinetic energies is plotted versus center-of-mass reaction angle, for projectile-like fragments from the reaction ^{209}Bi + ^{136}Xe at a laboratory energy of 940 MeV. The horizontal lines represent the fits to the angle-integrated Z distributions $d^2\sigma/dZdE$ (202).

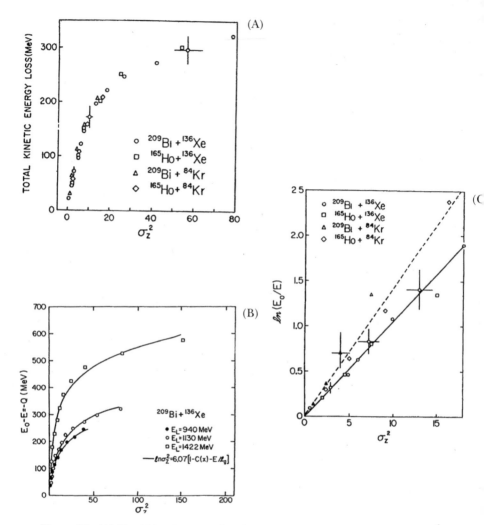

Figure 33. (A) Total kinetic energy loss in MeV as a function of the variance, σ_z^2, of the charge distribution of the projectile-like fragment. All values of the variance are determined from angle-integrated data (161). (B) Correlation between the variance σ_z^2 of the charge distribution and total kinetic energy loss for projectile-like fragments for the reaction $^{209}Bi + {}^{136}Xe$ at three laboratory energies (211). (C) Plot of $ln(E_o/E)$ as a function of the variance, σ_z^2, of the charge distribution of the projectile-like fragment (161). See text for definitions of E and E_o. The solid line is drawn through the xenon data. The dashed line is calculated relative to the solid line for the krypton-induced reactions by equation [26].

Assuming a Fokker-Planck equation to govern the exchange process, in the simple case of uncorrelated exchange of protons and neutrons, both fragment Z and A distributions are described in terms of Gaussians with first and second moments that increase linearly with reaction time t. It is further assumed that dissipation of kinetic energy T of relative motion occurs on the same time scale and is mediated by a classical force characterized by a force constant k such that (161),

$$-dT/dt = 2(k/\mu)T \qquad [25]$$

where μ is the reduced mass. With $d\sigma_z^2/dt = 2D_z$, one arrives at the relation (161),

$$ln\,(E/E_o) = [k/(\mu D_z)]\sigma_z^2 \qquad [26]$$

under the condition that the ratio $k/(\mu D_{zz})$ does not vary along the trajectory. Here, E_o and E are the initial and final kinetic energies as defined previously. Hence, a linear relation is predicted between $ln(E/E_o)$ and σ_z^2. Such a plot is shown in figure 33C where the Coulomb barrier is calculated at the strong absorption radius. Good agreement between experiment and theory is obtained for energy losses up to 200 MeV. However, such a simple theory neglects several factors including energy tied up in collective degrees of freedom.

If $k/(D_{zz})$ is the same for Kr- and Xe-induced reactions, the slope given by Equation [26] is slightly larger for Kr-induced reactions (compare the dashed and solid line in figure 33C). The data plotted in figure 33C support a rate of energy loss that depends on the square of the relative velocity or on the available energy. From the two slopes shown in figure 33C, an average value of $k/(D_{zz})=(0.9 \pm 0.3)$ x 10^{-43} MeV sec^{2}fm^{-2} is obtained. From the relationship between σ_z^2 and the interaction time for the ^{165}Ho + ^{84}Kr reaction, a value of the proton diffusion coefficient $D_{zz}= (0.7 \pm 0.3)$ x10^{22} sec^{-1} is deduced for the nonsticking model. The corresponding value of the frictional constant, k, is (0.6 ± 0.3) x 10^{-21} MeV sec fm^{-2}.

We studied the interrelation between energy dissipation, nucleon exchange, and the microscopic time scale of heavy-ion collisions for a number of reactions (162, 169, 170, 172, 174, 211, 215, 216) in terms of classical phenomenological models. In the mid-nineteen seventies, we wrote the first comprehensive review (172) of "Damped Heavy-Ion Reactions," which was published in the Annual Review of Nuclear Science in 1977. This was the second time that I had been invited to review a research field in the Annual Review of Nuclear Science. In 1972, I had been invited to review the subject of "Nuclear Level Densities" (133). The 1977 article was widely quoted and referenced (as was the 1972 article), and we received hundreds of requests for reprints. The major sections of this article were (1) Introduction, (2) Macroscopic and microscopic approaches to heavy-ion reactions, (3) Heavy-ion potentials and interaction radii, (4) Characteristic properties of strongly damped heavy-ion collisions, (5) Intercorrelation of experimental properties of damped collisions, (6) Emission of light particles and γ-rays in damped reactions, and (7) Summary. Only some three years had elapsed since our first publication on this subject and our 1977 review had already more than 300 references, attesting to the high level of research activity in this new field.

During 1977, I presented eleven invited lectures at various universities and conferences. One of these lectures (169) presented at the Conference in Fall Creek Falls State Park in Tennessee was organized by the Oak Ridge National Laboratory and attended widely by scientists from many countries. The abstract of this paper included the following: "Correlations of experimental observables with kinetic energy loss and fragment mass for damped heavy-ion reactions are emphasized in this paper. Angular-momentum-dependent interaction times are deduced giving a timescale for the evaluation of nuclear diffusion coefficients. The energy dissipated per nucleon exchange in krypton- and xenon-induced reactions is shown to decrease with increasing kinetic energy loss. These results are compared with predictions of a one-body dissipation mechanism and microscopic transport theory for heavy ion collisions."

In addition to krypton and xenon projectiles, we also used ^{56}Fe and ^{40}Ca projectiles in our various studies of strongly damped collisions. These lighter projectiles have some advantages in some types of experiments. For example, the N/Z ratios are very different for the two nuclei in the entrance channel, when using targets such as ^{165}Ho, ^{209}Bi, and ^{238}U. I will first discuss neutron emission in the ^{165}Ho + ^{56}Fe reaction at a laboratory bombarding energy of 476 MeV (185, 188, 193).

After the breakup of the intermediate complex produced in a damped collision into two excited final fragments, the communication between the two fragments is terminated; and the excitation energy acquired by the two fragments leads to the evaporation of light particles, as well as gamma rays during the final stages of the de-excitation cascade. Although emitted only in a secondary process, de-excitation particles from the final fragments carry important information on the damped-reaction mechanism, in particular on the energy-dissipation processes involved. Of considerable interest in this respect is the key question of the present investigation: Whether or not statistical equilibration of the excitation-energy degree of freedom is achieved during the short interaction time encountered in a damped collision. Only if such an equilibration has occurred does one expect the final fragments to acquire the same nuclear temperature.

Since the particle-evaporation spectra reflect the fragment nuclear temperature, it is possible to investigate the energy-equilibration processes pertaining to damped collisions by measuring the secondary particles. The study of neutron emission is advantageous because of the absence of Coulomb effects. A test of the degree of excitation-energy equilibration achieved in a damped collision is expected to be more conclusive for a system where the projectile and target masses are significantly different, as they are for the ^{165}Ho + ^{56}Fe reaction.

Identification of the dominant emission sources for neutron production in damped and fusion-fission-like reactions relies on the correlation of particle energies and emission angles with the velocities of the emitters. The usually quite low average neutron energies for statistical emission from a nucleus of temperature τ given by $<E_n>=(3/2)\tau$, ensure an efficient kinematical focusing of the majority of neutrons with low rest-frame energies into a relatively narrow cone centered around the direction of flight of the emitting nucleus.

The experimental neutron multiplicities for the light and heavy fragments are plotted in figure 34B versus the measured atomic number of the detected light fragment and that of the correlated heavy fragment. The multiplicities increase approximately linearly with increasing atomic number or mass of the fragment. The discontinuity at $Z=46$ is due to charge-particle emission from the fragments. In figure 34A, the ratio of the neutron multiplicities from the heavy to the light fragment is plotted and compared with the fragment-mass ratio (solid line). figure 34C demonstrates the typical evaporation shapes that neutron energy spectra exhibit in the rest frames of the light and heavy fragments. Their exponential shapes indicate equal nuclear temperatures despite different fragment masses and, hence, suggest thermalization of the dissipated kinetic energy within the reaction time. An estimated $(2–4) \times 10^{-22}$ seconds is needed in this reaction to dissipate and equilibrate an energy of 60 MeV. Such observations suggest a very intimate contact of the interacting nuclei forming an intermediate dinuclear system, and imply a rapid response of the intrinsic motion of the changing collective variables.

The experimental dependence of the nuclear temperature on the energy loss is shown in figure 34D. One observes that the temperatures deduced from the center-of-mass neutron-energy spectra are, within the experimental uncertainty, the same for both fragments already at the smallest energy loss, where the interaction time is approximately 4×10^{-22} second. Furthermore, the data follow closely the theoretical curve representing the Fermi-gas relation between the temperature and excitation energy assuming that the energy loss is converted into intrinsic excitation of the fragments. We conclude that the equal temperatures for light and heavy fragments shows that complete equilibration of the excitation energy is reached in damped heavy-ion collisions.

Structure in the energy spectra of specific final fragments produced in symmetric heavy-ion damped collisions (187) is shown to correlate with statistical evaporation processes from primary fragments. Calculations are described and applied for ^{41}Ca produced by the ^{40}Ca + ^{40}Ca reaction at a bombarding energy of 400 MeV. Contrary to exotic explanations of others, we showed that structure of the type found experimentally is produced by statistical processes following the production of primary fragments by the well-known nucleon exchange process.

Studies of product mass and charge distributions have demonstrated that the nucleon-exchange process in damped heavy-ion collisions is strongly influenced by the N/Z ratio of the composite system. Consequently, in reactions

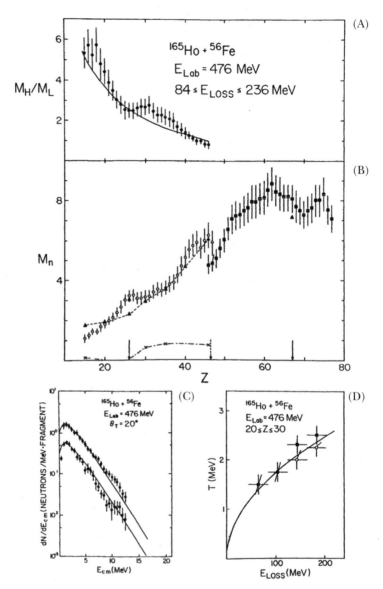

Figure 34. (A) The ratio of neutron multiplicities from the heavy and light fragment is plotted versus the charge of the light fragment. The solid line is the mass ratio of the heavy to the light fragment (188). (B) Neutron multiplicities of the light (open) and heavy (closed) fragments plotted versus the charge of the fragment (188). (C) Neutron-energy spectra in the rest frames of the light (open) and heavy (closed) fragments in damped reactions (207). (D) Nuclear temperature as a function of the energy loss for the light (open) and heavy (closed) fragments. The solid curve represents the calculated nuclear temperature (188).

between projectiles and targets, which differ significantly in their N/Z ratios, the lighter products tend to be neutron rich. This feature of damped collisions was utilized (204) in some ^{56}Fe bombardments of ^{209}Bi and ^{238}U to produce previously unobserved nuclei of ^{52,53}Sc, ^{54,55}Ti, ^{56}V, and ^{58,59}Cr as well as the tentative identification of six other nuclei.

The ^{165}Ho + ^{56}Fe system is a projectile/target combination that enables the study of all facets of heavy-ion reaction mechanisms, ranging from elastic scattering to fusion-fission. This system was studied at a laboratory bombarding energy of 462 MeV (212, 217). The distribution of the double differential cross section with respect to final kinetic energy E and fragment Z is shown in figure 35A in terms of a contour plot, integrated over all angles. Starting from the iron projectile Z=26, the cross section develops a ridge decreasing in Z and broadening as kinetic energy is dissipated. Also observable is a small peak near Z=47 where fusion-fission fragments are expected.

The angular distribution of the projectile-like damped fragments is plotted in figure 35B as $d^2\sigma/d\theta_{cm}dE$ versus angle for 20 MeV wide energy bins. At low energy losses the data show a pronounced peak near the quarter-point angle. At the highest energy losses, the double differential cross sections level out to a near constant value, which is equivalent to a $1/\sin\theta$ dependence. The narrow angular distributions observed for small energy losses are characteristic of short interaction times. As energy is dissipated, broader angular distributions result, suggesting a steady increase of interaction time with energy loss.

In figure 35C are shown results of Gaussian fits to fragment Z distributions for different scattering angles and various final energies. Although the width (FWHM) of the charge distribution increases with decreasing final kinetic energy, it is nearly independent of angle for a given energy bin. Hence, the reaction angle is not a relevant parameter to describe the evaluation of this reaction.

In figure 35D, the angular distribution for fusion-fission events is shown as $d\sigma/d\theta_{cm}$ versus θ_{cm}. The data displayed in figure 35D are consistent with a $1/\sin\theta_{cm}$ angular dependence, indicated by the horizontal line in this figure. The fusion-fission cross section is 720 millibarns at a laboratory energy of 462 MeV, indicating the angular momenta $l \leq 125\hbar$ contribute to the fusion-fission process.

Alpha-particle emission in coincidence with damped reaction fragments is studied for the ^{165}Ho + ^{56}Fe reaction at a laboratory energy of 465 MeV (222). Statistical emission from the two fully accelerated strongly damped reaction fragments accounts for the dominant sources of the alpha particles. The alpha-particle multiplicity M_α depends strongly on the fragment excitation energy and atomic number, with an average multiplicity $<M_\alpha>=0.2$ for iron-like fragments from damped reaction events with energy losses greater than 100 MeV. Temperatures derived from the alpha-particle and neutron spectra are in good agreement.

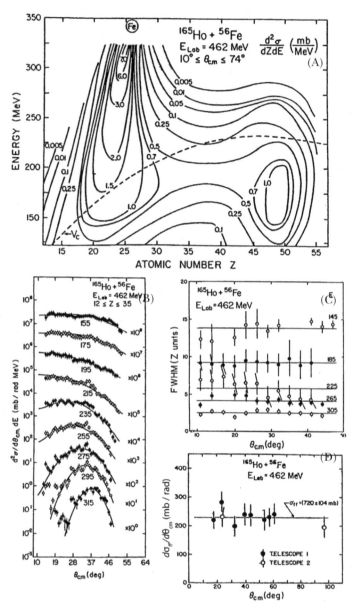

Figure 35. (A) Contour plot of the total energy versus Z integrated over angle. Full lines are given in units of $d^2\sigma/dZdE$. The dashed line represents the Coulomb barrier for touching spheres for different fragmentations (212). (B) Angular distributions plotted as the double differential cross sections $d^2\sigma/d\theta_{cm}dE$ versus scattering angle, as a function of the total-kinetic energy. The energy bins are 20 MeV wide. The data are integrated over all damped fragment charges, excluding fusion-fission (212). (C) Widths of Z distribution versus angle shown for five energy bins, 40 MeV wide. The horizontal lines are the average widths for each energy window (212). (D) Differential cross section $d\sigma_{ff}/d\theta_{cm}$ versus angle for fusion-fission events (212).

Detailed studies were also made of the multiplicity, energy, and angular distributions of alpha particles emitted in coincidence with fission fragments from the ^{165}Ho + ^{56}Fe reaction at a laboratory energy of 465 MeV (219). Approximately 90 percent of the alpha particles are evaporated from the two fully accelerated fission fragments. The remaining alpha particles have a strongly enhanced emission probability in the direction perpendicular to the scission axis, suggesting the contact or neck region between the fragments as the source of the particles.

Yields were measured with discrete Z and A resolution for projectile-like fragments produced in damped collisions between ^{56}Fe projectiles, with a laboratory energy of 465 MeV, and ^{56}Fe, ^{165}Ho, and ^{209}Bi targets (186). From the measured values of Z and A, corrected for neutron evaporation, it is possible to deduce the dependence of $\overline{A}/\overline{Z}$ on the energy loss, as shown in figure 36A. The data support a nucleon-exchange mechanism in which there is an increase in $\overline{A}/\overline{Z}$ that begins to saturate at the highest energy losses at a value of approximately 2.3 for the projectile-like fragments for both the ^{165}Ho + ^{56}Fe and ^{209}Bi + ^{56}Fe reactions. The dependence of the ratio of variances, σ_A^2/σ_Z^2, on energy loss is plotted in figure 36B for the ^{56}Fe + ^{56}Fe, ^{165}Ho + ^{56}Fe, and ^{209}Bi + ^{56}Fe systems. The lower and upper dashed lines represent the A/Z and the (A/Z)^2 values of the composite systems, respectively. It can be seen in figure 36B that the limit of correlated exchange is reached for the Fe + Fe system at an energy loss of about 75 MeV, whereas the asymmetric systems require a larger energy loss to reach that point.

Yields for specific Z and A were measured for projectile-like fragments produced in the reaction of 465-MeV ^{56}Fe ions with targets of ^{56}Fe, ^{165}Ho, and ^{238}U (203, 220). Variances of the isobaric charge distributions, $\sigma_Z^2(A)$, reveal a saturation value of $\sigma_Z^2(A) \approx 0.8$, reached within the first 30 to 50 MeV of energy loss depending on the target. Variances of the isotopic mass distributions saturate at a value of $\sigma_A^2(Z) \approx 2$–4, which is reached after about 60 to 80 MeV of energy loss. The neutron and proton variances indicate an early dominant neutron flow followed by a constant relative neutron and proton exchange rate at energy losses above about 30 MeV (220). The degree of correlation of the neutron-proton distributions increases with energy loss and is found to be strongly related to the potential energy surface.

In a phenomenological approach described in publications (170, 174, 180), use was made of the microscopic time scale provided by the nucleon exchange mechanism to derive a dissipation rate (195, 197):

$$-dE/dN_{ex} \approx (m/\mu)\alpha E \qquad [27]$$

with respect to the number N_{ex} of nucleons exchanged. Here, E=E$_{cm}$–V$_{coul}$–E$_{loss}$ is the available relative kinetic energy above the Coulomb barrier V$_{coul}$, m is the nucleon mass, and μ is the reduced mass of the dinuclear system. The coefficient α conveys information on the character of the dissipation mechanism.

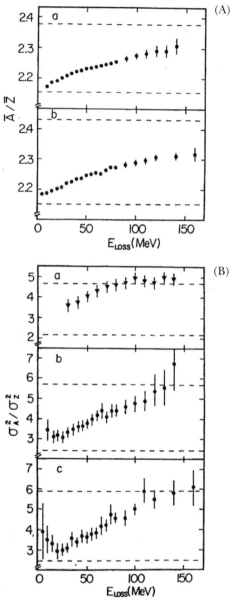

Figure 36. (A) Ratios of $\overline{A}/\overline{Z}$ as a function of energy loss for the (a) ^{165}Ho and (b) ^{209}Bi + ^{56}Fe systems. The dashed lines give A/Z of the projectile and composite systems (186). (B) Ratios of σ_A^2/σ_Z^2 as a function of energy loss for the (a) ^{56}Fe, (b) ^{165}Ho, (c) ^{209}Bi + ^{56}Fe systems. Data points in (a) (triangles) represent the measured values, in (b) and (c) (circles) values corrected for neutron evaporation. The dashed lines represent A/Z and (A/Z)^2 values of the composite system (186).

Under the assumption that α remains constant during a given collision trajectory, Equation [27] may be integrated yielding

$$E_{loss} \approx (E_{cm} - V_{coul})\{1 - \exp[-(m/\mu)\alpha N_{ex}]\} \qquad [28]$$

Figure 37A shows the resulting fits of Equation [28] to the experimental data for the reaction ^{209}Bi + ^{136}Xe at 1130 and 940 MeV. As can be seen, an adequate representation of the data is provided by Equation [28] up to energy losses close to the initially available energies $E_o = E_{cm} - V_{coul}$ indicated by arrows. However, it is observed that α decreases from 3.2 to 1.4 as the bombarding energy is increased from 940 to 1130 MeV. A similar behavior is borne out by all other reactions studied, as demonstrated in figure 37B exhibiting the dependence of α on the approach energy per nucleon. Deviations from the average trend are observed for the asymmetric systems associated with Fe projectiles as well as the ^{238}U + ^{238}U reaction. The dependence of α on both the bombarding energy and projectile-target combination appears to be inconsistent with models based principally on classical kinematic considerations. Thus, it was then conjectured that the insufficiency of the classical model to describe the data is due to its neglect of the quantal character of the nucleon exchange and dissipation mechanisms.

A recent model of Randrup was then applied to describe the energy dissipation associated with the exchange of nucleons between two Fermi-Dirac gases in slow relative motion characterized by a relative velocity $\dot{U}$. In this model the particle-number dispersion depends explicitly on the correlations present, such as those imposed by Fermi-Dirac statistics of the nucleons. The rate of growth of σ_A^2 is given by

$$d\sigma_A^2/dt \approx 2\tau^* N'(\varepsilon_F) \qquad [29]$$

where $\tau^* = \left\langle \frac{1}{2}w\coth(w/2\tau)\right\rangle_F$ is a measure of the energy interval around the Fermi level contributing to exchange processes. The quantity $N'(\varepsilon_F)$ is the differential current of nucleons exchanged between the gases calculated with neglect of the Pauli blocking effect. The appearance of τ^* in Equation [29] ensures that proper account is taken of the quantum statistics at all temperatures.

For symmetric systems and peripheral collisions,

$$\alpha = (\mu/mE)dE_{loss}/d\sigma_A^2 \approx T_F/2\tau^* \qquad [30]$$

where T_F is the Fermi energy. In a classical treatment neglecting the Pauli blocking effect, $\tau^* \to T_F/2$ and yields $\alpha \approx 1$ (see figure 37B). The above simple estimate suggests that α should typically be substantially larger than unity and decrease as the bombarding energy is increased. Furthermore, according to Equation [30], a certain degree of universality is expected for α. It is striking to observe that

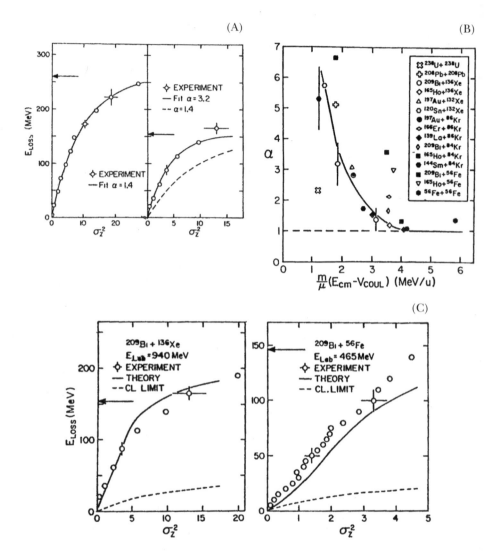

Figure 37. (A) Correlation between energy loss and variance of the fragment-Z distribution for the reaction ^{209}Bi + ^{136}Xe at a laboratory energy of 1130 MeV (left) and 940 MeV (right). Arrows denote initially available kinetic energies. The curves represent predictions according to Equation [28] with indicated α values (195). (B) Values of the parameter α obtained from fits of equation [28] to available data plotted versus the initially available energy per nucleon above the Coulomb barrier (195). (C) Comparison of model predictions for the correlation $E_{Loss}(\sigma_Z^2)$ with data for the reactions ^{209}Bi + ^{136}Xe (left) and ^{209}Bi + ^{56}Fe (right). The dashed curves represent the classical limits of the full calculations (solid curves) (195).

these general features, absent in a classical picture, are indeed borne out by the experimental results shown in figure 37B.

Since the estimate represented by Equation [30] relies on a number of idealizations, we chose a more refined approach by performing dynamical calculations of collision trajectories in a coordinate space including the fragment-mass and -charge asymmetries. The dinuclear complex is parameterized by two spherical nuclei joined by a cylindrical neck. Conservative forces are calculated from droplet-model masses, the Coulomb repulsion, and the surface and proximity energies of the neck region. Energy dissipation is provided by the nucleon-exchange mechanism, together with the damping due to the neck motion approximated by a wall-type dissipation formula. Inertial forces are calculated for two sharp rigid spheres. Energy loss and the accumulated variances σ_Z^2 and σ_A^2 are obtained from integrating along the trajectory, the dissipation function, and the diffusion coefficients, respectively, as given by the model.

Typical results of these calculations are compared with experimental results in figure 37C for two reactions that are associated with α values far in excess of the classical limit. Good agreement is obtained between the dynamical calculations and the data. As is illustrated by the dashed curves in figure 37C representing the dynamical calculations in the classical limit, the Pauli principle is essential to the good agreement between data and the quantal model. In summary, the good agreement between data and model predictions demonstrates that energy dissipation in damped reactions can be consistently understood in terms of a nucleon exchange mechanism in which the Pauli exclusion principle plays a crucial role.

Experimental relations between mass, charge, and energy loss of fragments from damped reactions were analyzed (206) in terms of the independent-particle transport model (195) described previously. Results of the model calculations (206, 207) are compared with experimental data (186, 203) of ^{56}Fe-induced reactions in figure 38. These data were taken in-beam and with better than unit mass and charge resolution, avoiding resolution effects obscuring other published data. As can be seen in figure 38A, ratios of average values $\bar{A}/\bar{Z}$ and σ_A^2/σ_Z^2 of the iron-like fragments from the reaction ^{165}Ho + ^{56}Fe and their variation with energy loss are well represented by the model calculations employing no adjustable parameters. Predicted ratios σ_A^2/σ_Z^2 are seen to decrease initially with increasing energy loss. In the hypothetical case of constant drift coefficients, the values σ_A^2/σ_Z^2 would decrease monotonically with increasing energy loss and approach a limit of approximately $\bar{A}/\bar{Z}$. Hence, the increase of the ratio σ_A^2/σ_Z^2 for energy losses beyond about 20 MeV can be interpreted as being due to strong interdependences between proton and neutron exchange processes, enforced by the dynamical driving forces.

Isotopic variances σ_A^2 ($Z = const$) of the fragment mass distributions are displayed in figure 38B for the above reaction, for several values of Z. An impressive description of the data by the model is observed. As predicted, these variances do not depend markedly on the Z values of the fragment chosen.

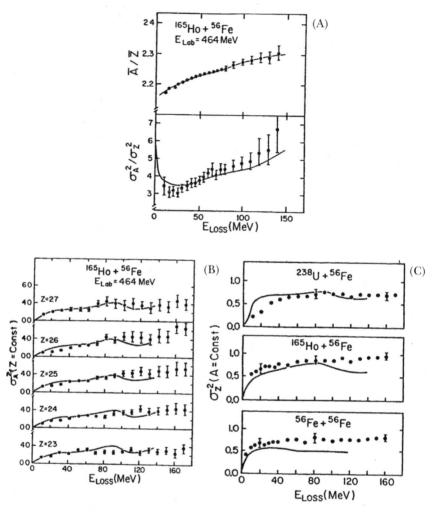

Figure 38. (A) Energy-loss dependence of experimental values (points) for ratios $\bar{A}/\bar{Z}$ of centroids (top) and ratios σ_A^2/σ_Z^2 of variances (bottom) of fragment A and Z distributions for the reaction $^{165}Ho + ^{56}Fe$, corrected for neutron evaporation of the primary fragments. Curves represent theoretical predictions (206). (B) Energy-loss dependence of experimental isotopic variances σ_A^2 (Z) (points) of mass distributions for fragments of a particular Z value. Variances are corrected for neutron evaporation. Curves represent theoretical predictions (206). (C) Energy-loss dependence of experimental isobaric variances σ_Z^2 (A) (points) of charge distributions for fragments of a particular A value. The $^{56}Fe + ^{56}Fe$ data are uncorrected for light-particle evaporation. Experimental values are averages over individual fragment masses. Curves represent theoretical predictions (206).

Isobaric charge variances σ_Z^2 (A=const) averaged over all individual A values are plotted in figure 38C versus energy loss, for three ^{56}Fe-induced reactions. A satisfactory agreement between data and calculations is noted for the ^{238}U + ^{56}Fe and ^{165}Ho + ^{56}Fe reactions, up to an energy loss of about 100 MeV.

In summary, predictions of a transport model based on statistical exchange of individual nucleons with Pauli blocking are found to be in agreement with experimental fragment A and Z distributions from damped reactions. They also give a satisfactory account of data on other reaction properties such as energy losses and deflection functions. The observed variations of $\bar{A}/\bar{Z}$ and σ_A^2/σ_Z^2 ratios with energy loss are understood to reflect the strong interdependence between neutron and proton exchange processes imposed by the dynamical driving forces.

Consistently, after an initial increase with energy loss, predicted isotopic and isobaric charge variances exhibit a saturation behavior in the energy loss range where this correlation is observed to be important for the ratios σ_A^2/σ_Z^2. The predicted drop in the variances for larger values of energy loss may not lend itself to a test by the data, before de-excitation processes of the primary fragments are more fully understood. The success of the model employed lends strong support to the predominance of incoherent nucleon exchange in the dynamics of damped nuclear reactions.

Extensive tables of reaction parameters for heavy-ion collisions were tabulated and published (205). This 230-page publication allows quick reference to pertinent reaction parameters for 432 projectile-target combinations for a bombarding energy range 1 to 50 MeV/nucleon. The table entries were selected with the intention of providing experimentalists working in the fields of heavy-ion-induced fusion and damped reactions with useful parameters that characterize the main features of the angular and energy distributions of the reactions products.

A large area position sensitive ΔE–E detector for energetic light charged particles is described (208). A single wire gas proportional counter is used as the ΔE detector, with stop detectors consisting of three 19 millimeters thick NE102 scintillators coupled to RCA 8053 photomultipliers. The total active area of the ΔE–E detector is 43 square centimeters. The detector is suitable for alpha particles with energies between 6.5 and 180 MeV. The detector is compact enough to fit into a 30-inch scattering chamber.

3. Heavy-Ion Fusion

The deeply penetrating heavy-ion collisions may lead to fusion-fission-like and/ or compound nucleus reactions (see the schematic classification of heavy-ion reactions shown in figure 28). These reactions are characterized by complete loss of projectile and target identities and complete damping of kinetic energy.

Lighter-mass systems yield deep interpenetration more readily. The total heavy-ion fission cross section is obtained by summing the partial cross sections of the compound nucleus (CN) and the fusion-fission-like (FL) processes.

The measurement of heavy-ion fusion excitation functions offers the possibility of insight into the form of the internuclear conservative force, the magnitude and mechanisms of nuclear friction, and the limits of rotational stability of heavy nuclei. Analyses of experimental fusion excitation functions on the basis of simple friction-free models assume that a sufficient condition for the fusion of the target and projectile is either that a conservative potential barrier is crossed, or that a critical separation distance is reached.

The well-known damped-reaction process described in the previous section gives strong evidence for friction. Hence, we proposed a dynamical model for heavy-ion fusion based on the proximity nuclear potential and one-body friction (178, 183, 184, 189, 191, 194, 209, 223, 224). Experimental data on heavy-ion fusion provide an important source of information about the radial dependence of the nuclear conservative and dissipative forces. Our trajectory calculations make use of the proximity nuclear potential of Blocki et al. designated by Φ and shown in figure 39 (solid line) as a function of the separation distance ζ (in fermis) of the surfaces of the target and projectile nuclei. We compared this universal proximity potential with experimental elastic scattering and inelastic reaction data in figure 39 (171, 173, 182). The analysis of elastic scattering data produces the circles and the analysis of the inelastic reaction data produces the triangles in figure 39. The open triangles are based on excitation-function measurements of fusion cross sections by counter-telescope measurements of evaporation residuals and/or fission fragments, while the filled triangles rely on excitation functions based on summing measured partial fusion cross sections. The reaction data test the proximity potential at much smaller surface separation distances then does elastic scattering.

The experimental values of the proximity potential $\Phi_{\exp}(\zeta_B)$ at the fusion barrier are given by

$$\Phi_{\exp}(\zeta_B) = V_N(R_B)\{4\pi\gamma[C_T C_P / (C_T + C_P)]b\}^{-1} \tag{31}$$

where

$$\gamma = 0.9517\{1 - 1.7826[(N_T + N_P - Z_T - Z_P)/(A_T + A_P)]^2\},$$
$$C_i = R_i[1 - (b/R_i)^2 + ...],$$
$$R_i = 1.28 A_i^{1/3} - 0.76 + 0.8A_i^{-1/3},$$
$$\zeta_B = R_B - C_T - C_P$$

In equation 31, the quantities C_i, R_i, and b (b=1) are in units of fermi, $V_N(R_B)$ is in units of MeV and γ in units of MeV fm^{-2}. The experimental values of the dimensionless proximity potential, calculated from inelastic reaction data by way of Equation 31, are compared with the theoretical proximity potential in figure 39. Although the experimental values of the proximity potential derived from inelastic reaction data show considerable spread around the theoretical values, there is a qualitative agreement between experiment and theory down to a surface separation of one fermi (see figure 39).

One-dimensional classical dynamical models of fusion (178, 191, 223), which explicitly consider dissipative forces, have been successful in reproducing a considerable amount of fusion data over wide mass and energy ranges. Insofar that the model is based on average properties of the nuclear interactions, it can only be expected to reproduce general trends of fusion excitation functions and not necessarily give precise agreement with data in certain cases that may well reflect the individual structure of the particular nuclei involved.

In our model, four equations of motion (see, e.g., reference 191) of the two interacting heavy ions, described by four dynamical variables r, θ, θ_T, and θ_p, are solved numerically. The trajectory for each value of the angular momentum is followed as a function of time in order to determine whether the ions are trapped behind a potential barrier (leading to fusion) or escape over the barrier at a reduced energy (leading to a damped reaction). The first two degrees of freedom, r and θ, define the radial separation and angular orientation of the two-nuclear system and the second two degrees of freedom, θ_T, and θ_p, are the angles specifying the orientation of the target and projectile, respectively. For the conservative potentials, use is made in our trajectory calculations of the Bondorf Coulomb potential and the nuclear proximity potential of Blocki, Randrup et al. The effect that variations of those potentials have on predicted fusion cross sections was investigated (191). The radial and tangential frictional form factors needed to solve the equations of motion are calculated from the one-body frictional model of Randrup. All the nuclear-radius parameters of the model have been taken from liquid-drop-model systematics. All the quantities needed to solve the classical dynamical model are fixed leaving no adjustable parameters.

The results of the one-dimensional classical dynamical model have been compared with a wide variety of experimental data (178, 183, 184, 189, 191, 194, 209, 223, 224). Two of these publications deserve special mention. In publication (191), extensive calculations were performed to test the sensitivity of the calculated fusion cross sections to a number of parameters, including the radial dependence of the Coulomb and nuclear potentials, the radial and tangential friction form factors as well as the projectile and target radii. The theoretical excitation functions for the lighter heavy-ion systems are rather insensitive to changes in either the conservative or dissipative forces. The calculations show that tangential friction sufficient to produce the rolling condition is necessary to explain the magnitude of the fusion cross sections at high energies, which are also sensitive to the magnitude of the radial friction component. This is in contrast to the fusion cross sections at low energies that are determined by the

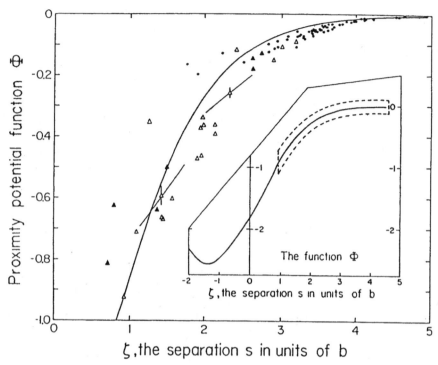

Figure 39. A comparison of the universal proximity potential of Blocki et al. (solid line) with experimental data (points) as a function of the surface separation of two nuclei in units of fermi (173).

nuclear potential at larger separations, and to a lesser extent by tangential friction. The low energy fusion data are most sensitive to nuclear radii. The fusion cross sections at high energies, especially for heavy systems, are the most sensitive to the assumed forces. However, even for these cases the effects of the conservative and dissipative forces are similar and difficult to separate.

A second article of note was a review article (223) entitled "Fusion Reactions between Heavy Nuclei." This article marked the third time that we had been invited to review a major field of nuclear science for the Annual Review of Nuclear and Particle Science. The table of information on heavy-ion reactions in the article contains 385 entries.

Some examples of the comparison between theoretical and experimental heavy-ion fusion cross sections are shown in figure 40. In Figures 40A and 40B, the fusion cross section is plotted as a function of the inverse in the center-of-mass energy for the reactions ^{112,116,120,124}Sn + ^{35}Cl and ^{109}Ag + ^{40}Ar, respectively. In these reactions, friction becomes important only at the higher energies. In Figures 40C and 40D, the abscissa is in the center-of-mass energy in order to display the higher energies more effectively. In figure 40C, one observes a close agreement between theory and experiment as the fusion cross section increases initially with energy, reaches a maximum, and then decreases smoothly with further increase in energy.

It is well known that the cross sections for fission-like fragments produced in heavy-ion reactions on heavy targets may exceed considerably the cross-section limits expected from the rotating liquid-drop model which considers the stability of a rotating compound nucleus against symmetric deformations and fission (209). These experimental results, therefore, suggest the existence of a long-lived asymmetric dinuclear system trapped with high angular momentum that can be sustained by the system. These trapped systems, although ultimately unstable, have interaction times sufficiently long to allow considerable relaxation of even slowly equilibrating degrees of freedom, such as mass asymmetry. Because of the long interaction times, the fragments resulting from these trapped systems exhibit a $1/\sin\theta_{cm}$ angular distribution (see figure 35D) and have masses and kinetic energies essentially consistent with a fusion-fission reaction process.

The ^{58}Fe + ^{208}Pb reaction shown in figure 40D is such a system. As can be seen in this figure, the one-dimensional classical model (dashed line) over estimates the fusion cross section for such a reaction with large values of Z_P and Z_T. The solid line in figure 40D is calculated with a dynamical theory of Swiatecki, which includes neck and mass asymmetry degrees of freedom, in addition to the fragment separation degree of freedom considered in the simple classical dynamical theory.

In summary, the one-dimensional classical dynamical model reproduces rather well the shape and magnitude of fusion excitation functions for light and intermediate systems. This supports the view that, in general, fusion is limited by entrance channel restrictions. The simple one-dimensional model fails to reproduce experimental fusion excitation functions for systems composed of large values of both Z_P and Z_T. For these heavy systems, mainly two effects appear to be responsible for the limitation of fusion. On the one hand, Coulomb and centrifugal forces are sufficiently repulsive to prevent configurations of tangent nuclei to evolve toward fusion. In addition, for heavy systems, the potential energy landscape at large nuclear overlaps is significantly altered by the accessibility of other degrees of freedom such as associated with the formation of a neck, by which a system may escape from the fusion path. This escape into another dimension can be prevented by enforcing deeper interpenetrations by sufficiently large kinetic energies.

C. Actinide Muonic Atoms

In the late 1960s and early 1970s, I participated in several discussion groups and study conferences to plan the research use of a new national meson physics facility to be built at Los Alamos. During August 14–23, 1968, I was invited to the first

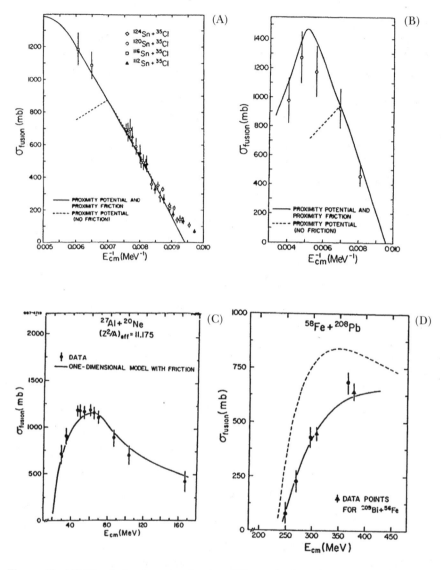

Figure 40. (A) The fusion cross section in millibarns is plotted versus the inverse of the center-of-mass energy. The theoretical lines are for the $^{116}Sn + {}^{35}Cl$ reaction (178). (B) Same as (A) except for the $^{109}Ag + {}^{40}Ar$ reaction (178). (C) The fusion cross section in millibarns is plotted versus the center-of-mass energy (224). (D) Fusion-fission cross section in millibarns is plotted versus the center-of-mass energy. The dashed line is calculated with the one-dimensional classical dynamical model. The solid line is calculated with a dynamical theory of Swiatecki, which includes neck and mass asymmetry degrees of freedom (209).

small informal conference at Los Alamos to prepare for the nuclear chemical research at the facility. It was anticipated that by early 1972 (actually ready for use in 1973), Los Alamos would have a high-flux proton linear accelerator producing one milliampere of protons (6.24×10^{15} per second) at 800 MeV. These new facilities would exist for research use with these protons and with the associated secondary beams of intrinsic fluxes of mesons and neutrons.

Our experiments at the Los Alamos Meson Physics Facility (LAMPF), a half-mile-long accelerator sitting on a mesa, were performed using the stopped-muon channel. The muons are produced from the decay of pi mesons. The secondary beam of pi mesons at LAMPF, the most intense of such beams available, is produced by the intense beam of 800-MeV protons striking light-element targets. The negative muons (μ^-) from the decay of the secondary pions are magnetically guided down the facilities' muon channel to our actinide targets. The negative muon is captured in an outer orbit of the actinide atom and cascades down to the lowest energy orbit (1s) in much less than a picosecond. The radii of muonic orbits are some 200 times smaller than the corresponding electron orbits due to the fact that the muon is approximately 200 times heavier than the electron. Hence, a muon in a 1s orbit of an actinide atom spends a considerable time in the nucleus and has a high probability of being captured.

Our experiments, designed to determine the muon capture rate in neutron-rich, actinide nuclei, measured the time difference between the arrival of a muon in the target and the appearance of any reaction product associated with the capture of the muon by the nucleus. Since muon capture excites a heavy nucleus on the average to approximately 15 MeV, various reaction products including neutrons, fission fragments, and gamma rays can be observed. In addition, it is possible to infer the muon-capture rate from the time distribution of electrons emitted in the leptonic decay of muons in the 1s state of an actinide muonic atom.

It was generally assumed that all these detection methods yield the same lifetimes for a particular muonic atom. In order to further test this assumption, we made measurements of the disappearance of muons from the 1s state of actinide muonic atoms by detecting muon-decay electrons (165), neutrons (181), and fission fragments (190, 198). The lifetime measured by detecting fission fragments emitted after muon capture were consistently shorter than lifetimes determined with any other method. For actinide nuclei, prompt neutron emission or fission may occur during the atomic cascade of a muon to its ground state due to a radiationless transfer of the muonic transition energy ($1p \rightarrow 1s$) to the nucleus. In the course of such a prompt fission process, the muon may become bound in an atomic orbital of one of the fission fragments and may later be captured by the fragment or undergo leptonic decay. In a singles experiment, electrons, neutrons, and gamma rays resulting from muonic fission fragments cannot be distinguished from these stemming from muon capture by actinide nuclei. Hence, the time distributions of electrons, neutrons, and gamma rays consist of a superposition of two exponential components. The characteristic lifetimes are 70 to 80 nanoseconds for actinide muonic atoms and about 130 nanoseconds for fission-fragment muonic atoms. The statistics of most experiments is not sufficient to allow a distinction of the various components of the measured time distributions such that least-square fits assuming a simple exponential decay law for these distributions will yield lifetimes that are too long.

In contrast, because muon capture by a fission fragment cannot induce fission of this fragment, time distributions of delayed fission fragments produced by muon capture are not disturbed by prompt fission events. There are, however, secondary effects that lead to insignificant contamination of these time distributions. For example, the radiationless atomic transition of a muon can cause the emission of one neutron from the target nucleus before subsequent fission of the daughter nucleus occurs after muon capture. This process has a probability similar to prompt fission, but since the lifetime difference between two neighboring isotopes is only about 1.5 nanoseconds for actinide nuclei, the effect of this admixture on the lifetimes is very small.

A schematic diagram of the experimental setup that we used to measure the lifetimes of actinide muonic atoms in the LAMPF muon channel is shown in figure 41A. A muon entering the counter telescope and stopped in the actinide target gives a start signal $123\overline{4}$, where counters 1, 2 and 3 detect a muon and counter 4 does not. With this start signal, the lifetime of the newly formed actinide muonic atom is measured by the delay times for detection of decay products. In the schematic shown, there is a fission chamber, between scintillation counters 3 and 4, for detection of fission fragments. A schematic drawing of the fission ionization chamber is also shown in figure 41B. The fission chamber contained three separate compartments, allowing simultaneous measurements of the lifetimes of three different actinide muonic atoms. Each of the three compartments of the chamber contained three titanium foils, on which the actinide targets (^{235}U, ^{238}U, ^{237}Np, ^{239}Pu, and ^{242}Pu) were deposited as 0.5 milligram per square centimeter thick oxide layers. Three foils carrying one particular nuclide formed an ionization chamber, and three such chambers constituted the whole fission chamber. All timing signals were processed in the same time-to-pulse-height converter and the same analog-to-digital converter (ADC), using routing signals to distinguish the ionization chamber that had fired. Time calibrations were performed frequently with several independent methods. In addition, only two nuclides were replaced at a time, so that the third nuclide remaining in the chamber served as an additional cross check for the consistency of the results for different runs.

A typical time spectrum, measured by detection of fission fragments, is shown in figure 41D. The data were analyzed with a least-square code and a fitting function representing an exponential distribution on a constant background. A trial fit assuming two exponentials failed to find a second lifetime component in the spectra, as was to be expected. Similar results were observed for the other

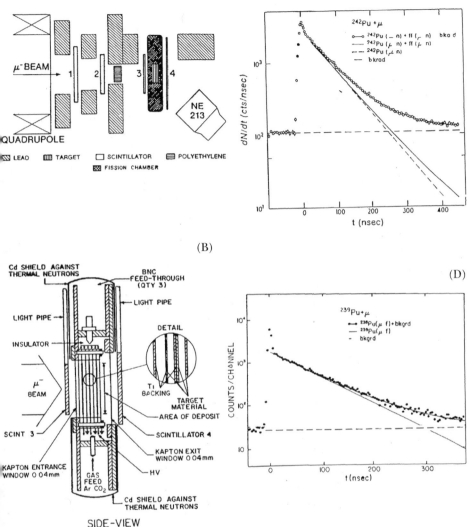

(A)

μ⁻ BEAM

QUADRUPOLE

▨ LEAD   ▥ TARGET   ☐ SCINTILLATOR   ▤ POLYETHYLENE
▨ FISSION CHAMBER

(C)

$^{242}Pu + \mu$

(B)

Cd SHIELD AGAINST THERMAL NEUTRONS

BNC FEED-THROUGH (QTY 3)

LIGHT PIPE

LIGHT PIPE

DETAIL

INSULATOR

μ⁻ BEAM

Ti BACKING

TARGET MATERIAL

AREA OF DEPOSIT

SCINT 3

SCINTILLATOR 4

KAPTON EXIT WINDOW 0 04mm

KAPTON ENTRANCE WINDOW 0 04mm

GAS FEED Ar CO₂

HV

Cd SHIELD AGAINST THERMAL NEUTRONS

SIDE-VIEW

(D)

$^{239}Pu + \mu$

Figure 41. (A) Schematic illustration of the experimental setup for measuring lifetimes of actinide muonic atoms by detecting fission fragments. The fission chamber is between scintillators 3 and 4 (190). (B) Schematic drawing of the fission ionization chamber. The insert shows the arrangement of the target foils (198). (C) Neutron-time distributions for muonic ^{242}Pu. The thin solid line includes contributions from muon capture in ^{242}Pu (dashed line) and fission fragments (ff) (181). (D) The experimental fission-fragment time distribution for muonic ^{239}Pu. The peak in the region near t = 0 corresponds to prompt fission events caused by radiationless muonic transitions. The horizontal dashed line represents a fit to the background to negative and large positive times. The curve drawn through the data points represents the sum of the exponential and the background distributions (198).

nuclei. The lifetimes of the muonic atoms, ^{235}U, ^{238}U, ^{237}Np, ^{239}Pu, and ^{242}Pu, deduced from a detection of delayed fission fragments, are 72.9 ± 0.9, 77.9 ± 0.5, 71.3 ± 0.9, 70.1 ± 0.7, and 75.4 ± 0.9 nanoseconds, respectively. As expected on theoretical grounds, the lifetimes of the higher mass (larger neutron number) isotopes of both uranium and plutonium are some 5 nanoseconds longer.

As mentioned previously, we also measured the lifetimes of a number of different actinide muonic atoms by detection of both muon-decay electrons (165) and muon-capture neutrons (181). In order to illustrate the small contamination introduced in the time spectra from these detection methods, I show a time distribution of neutrons from the decay of ^{242}Pu muonic atoms in figure 41C. The neutrons were detected with a large NE 213 scintillator coupled to a photomultiplier. A fast pulse-shape discrimination circuitry provided an efficient (> 98 percent) separation of neutrons from gamma rays. In this spectrum one sees a small contribution from fission fragments produced in prompt fission by radiationless muonic transitions.

The final experiment to be described in this section was designed to directly measure muon capture in secondary nuclei produced by prompt neutron emission and fission induced by radiationless transitions in the atomic cascade of muons bound to ^{237}Np, ^{239}Pu, and ^{242}Pu (190). Here I'll show results for ^{239}Pu muonic atoms as displayed in figure 42. The top part (A) of this figure shows the fission-fragment-neutron coincidence rate plotted in a two-dimensional diagram versus the times $t_{fission}$ and $t_{neutron}$ between a muon stop in the fission chamber and the detection of the resulting coincident fission fragment and neutron, respectively. Three ridges are clearly observable, extending from the prompt peak toward longer times. Ridge A is due to delayed muon-capture-induced fission of ^{239}Pu, where a fission fragment is detected in prompt coincidence with a neutron associated with de-excitation of either of the two fission fragments. Ridge B results from prompt emission of a neutron in a radiationless process followed by muon-capture-induced fission of the daughter isotope ^{238}Pu. Ridge C is due to prompt muon-induced fission of the target followed by muon capture in one of the fission fragments as identified by the appearance of a delayed neutron. This ridge is of particular interest.

Projecting ridges A and B onto the axis given by $t_{fission}$ and ridge C onto that corresponding to $t_{neutron}$, one obtains the time distributions of the three classes of events such as shown in figure 42B for a sample run on ^{239}Pu, where a constant background has been subtracted. The distributions reflect the lifetimes of muonic ^{239}Pu (A), of muonic ^{238}Pu (B), and of the muonic fission fragments (C). The data in ridge (B) illustrate for the first time a new technique for the measurement of the lifetime of actinide muonic atoms with one neutron removed from the target (although statistics are presently inadequate for a precise lifetime determination).

Whereas the experimental slopes of the time distributions are very similar for the target and its daughter produced by prompt neutron emission, the time

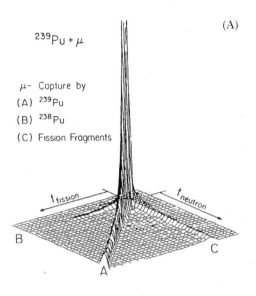

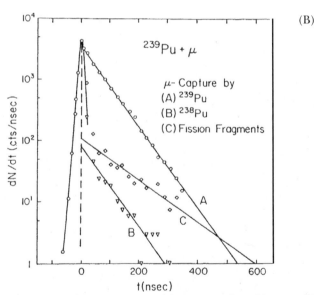

Figure 42. (A) Fission-fragment-neutron coincidence rate plotted in a two-dimensional diagram versus the times $t_{fission}$ and $t_{neutron}$ between the arrival of a muon in the fission chamber and the detection of the fission fragment and neutron, respectively (190) (see the text). (B) Decay curves of the three classes of events obtained by projecting the different ridges of the above figure onto the respective time axes (190) (see text).

distributions corresponding to the fission fragments (C) indicates a much lower rate of disappearance of the muon. The latter distributions were analyzed with a theoretical fit function including a random-time background and two exponentials corresponding to muon capture in the light and heavy fission fragments. From the data, however, only one exponential component was found with statistical significance, restricting a possible contribution from a second component to less than 10 percent. For all three targets, ^{237}Np, ^{239}Pu, and ^{242}Pu, the measured mean lifetime of the muonic fission fragments have nearly the same value of 130 nanoseconds. One concludes, therefore, that during the fission process, the muon attaches itself predominantly to the heavy-fission-fragment atom. This result shows that the rearrangement of the muonic orbit occurs adiabatically with respect to the fission process. If the nuclear fission mode is highly damped, the nuclear collective velocity from saddle point to scission is expected to be much smaller than typical velocities of the muon in its ground state. Such a situation, where the rearrangement of the muonic orbit occurs adiabatically with respect to the fission process, would result in a predominant capture of the muon by the heavy fission fragment.

D. Other Activities Including Lectures in China

Nuclear medicine, now an important component of current-day medicine, utilizes a number of radioactive nuclides in both the diagnostic and treatment aspects of medicine. Particular nuclides are sought for their suitable decay properties, such as half-life and type of radiations. Radioactive generator systems, consisting of both a radioactive parent and daughter, are particularly in demand. Such systems, with a long-lived parent and short-lived daughter, generate a useful short-lived radioactivity on demand without being near an accelerator or reactor. The parent radioactivity is stored on a resin column and the short-lived radioactive daughter eluded from the column as needed.

A new generator system was developed (199) using the ^{52}Fe $\rightarrow$ ^{52m}Mn parent-daughter pair. The ^{52}Fe with a half-life of 8.3 hours is isolated on an anion-exchange column, and ^{52m}Mn is eluted in hydrochloric acid. Breakthrough of the parent is less than 0.01 percent and the yield of the daughter is 75 percent. The 21.1-minute half-life of ^{52m}Mn is ideal for use in sequential studies, but is long enough to permit radiochemical manipulations to control biodistribution. Animal studies indicate that ^{52m}Mn is an ideal nuclide for myocardial imaging, combining rapid blood clearance and high concentration in the myocardium. An added advantage is that ^{52m}Mn decays 98 percent by positron emission and is useful for positron computer tomography. A variety of methods for production of ^{52}Fe were investigated (200).

On January 23, 1978, there was a symposium commemorating the twenty-fifth anniversary of the discovery of elements 99 and 100 held at the Lawrence Berkeley Laboratory. (I discussed the discovery of these elements in Chapter II of this manuscript.) At this symposium, I gave a lecture on the nuclear properties of these new elements, named einsteinium and fermium (175). I contributed an article on the newly discovered "Damped Nuclear Reactions" (see Chapter VIII and Subsection IX.B.2) for the 1978 edition of the McGraw-Hill Yearbook of Science and Technology (177). In early 1979, I was invited to Berlin to present a memorial lecture (192) honoring the great German scientists Lise Meitner and Otto Hahn, internationally known for their discovery of nuclear fission (the occasion was the fortieth anniversary of the discovery). An international symposium on the subject Continuum Spectra of Heavy Ion Reactions was held in San Antonio, Texas, on December 3–5, 1979. I was given the impossible task of summarizing all of the lectures of the symposium in thirty minutes. In my talk, published (196) in the proceedings, I chose to cover the invited and contributed papers under four categories: (1) properties of damped collisions at energies less than 5 MeV per nucleon above the Coulomb barrier; (2) emission of statistical light particles in damped collisions; (3) emission of fast (non-statistical) particles in heavy-ion collisions; and (4) heavy-ion fusion.

During the third Adriatic Europhysics Study conference on the Dynamics of Heavy-Ion Collisions held on May 25–30, 1981, on the Island of Hvar, Yugoslavia, I presented a lecture entitled "Heavy-Ion Fusion Revisited," which was published (209) in the conference proceedings. The conferees came from many countries and laboratories. Hvar is one of the most scenic locations for a scientific conference of all that I've attended; although Hercegnovi, the site of the Ninth Yugoslav Summer School, and also on the Adriatic, is a close rival. Five months later on October 26–30, I returned to the northern Adriatic region to present lectures at the Nuclear Physics Workshop, held at the International Center for Theoretical Physics in Trieste, Italy. These lectures entitled "Experimental Characteristics of Damped Reactions" were published (211) in a special edition of Nuclear Physics. The majority of those registered for this workshop were students from Europe, especially from the eastern countries.

In addition to the above lectures where manuscripts were required, I gave some eighty-five other invited lectures during this time period. Approximately thirty-five lectures were presented as colloquia at universities in the United States, ranging from the University of Colorado in the West to Yale and MIT on the East Coast to the University of Minnesota in the North and Texas in the South. Some ten lectures were given as conference talks or colloquia at national laboratories including the Lawrence Berkeley Laboratory on the West Coast to the Brookhaven National Laboratory on the East Coast. Some ten lectures were presented at national meetings of the American Physical Society (APS), the American Chemical Society (ACS), and the Gordon Research conferences. An additional five lectures were presented on a special ACS-sponsored lecture tour through South Carolina, Georgia, North Carolina, and Virginia in April 1978. These latter lectures discussed the heaviest elements in the periodic table and the possible existence of an island of superheavy elements.

The University of Rochester launched a new community lecture series in the fall of 1976 known as Wednesday Evenings at the University. Five such public lectures were given during the fall semester, and I presented one of them entitled Heavy-Ion-Reactions; A New Frontier in Nuclear Science. Popularizing basic science is not a talent that I claim to have mastered. I mentioned previously an experience I had testifying before the Joint Congressional Committee on Nuclear Energy. In popular lectures one attempts to give examples of classical phenomena, which are sometimes helpful but often misleading also in the quantum world. I began my lectures by showing illustrations of colliding water drops. However, making the transition to colliding nuclei, some twelve orders of magnitude smaller (10^{-12}) than water drops requires a good imagination. Although I received a number of favorable comments, I personally remained a skeptic about the degree of my success in the presentation of such a public lecture.

Including invited lectures at international conferences, I gave some twenty-five additional lectures abroad, mostly in Europe, during this nine-year period. Because of various other commitments, I had to decline a number of foreign invitations, including two requests from Professor G. N. Flerov to present papers at conferences in Dubna (USSR), a request to lecture at a conference in Japan, invitations to lecture at a summer school in Bangalore, India, and a workshop on nuclear physics in Bombay, as well as an all-expense-paid trip to South Africa to give the summary talk for an international conference.

Eight of the above foreign lectures were given in China during a one-month trip starting in early September, 1979. This trip had its origin in a book we published on Nuclear Fission in 1973 (see Subsection VII.B.4). One day a package of paperback books arrived with photographs and tables from this book and the text in Chinese. They had translated our book without our knowledge or permission! Later at a summer Gordon Conference, Professor Yang Fu-Chia (the surname appears first), Chairman of the Department of Nuclear Science of Fudan University in Shanghai at the time, informed my wife that they wanted to invite us to visit China as their guests (they inferred this was partial repayment for our loss of book royalties). An agreeable time period for our visit was eventually worked out. Their invitation included my giving three lectures at Fudan University, the leading educational institution in Shanghai, two lectures at the Institute of Atomic Energy outside Beijing, one lecture at the University of Beijing, one lecture at Lanchow University, and one lecture at the Institute of Atomic Energy in Lanchow.

After a couple of interesting days in Hong Kong, we entered the People's Republic of China on September 10, 1979, by way of a Hong Kong–Guangzhou train. Our first stop was Guangzhou (formerly Canton), an important industrial city in southern China and the chief trade outlet at that time. Our hotel was across the street from the Chinese Export Commodities Fairgrounds. At the time of our visit, China was essentially still a closed society and we saw few, if any, people from the West. We were provided a guide, by the name of Chao, who traveled with us during out entire trip. Chao, in his twenties, was an advanced student of English and already had mastered the language well. Our trip throughout China would not have been possible without him. In addition to Chao, we were provided a local guide at each new location we visited. This second guide, knowledgeable in the local area, came with an automobile and driver. Hence, we had first-class local transportation for viewing the local scenic sights, museums, and monuments. This was at a time when the transportation on the city streets was principally by bicycle and horse-drawn wagons. Major trips were taken either by train or airplane. During air travel, instead of supplying air conditioning, a fancy hand-controlled fan was supplied free on entering the airplane.

Our second stop was Guilin, known for its unusual scenery, especially the karst formations. These are irregular limestone regions with sinks, underground streams, and caverns. Millions of years of erosion by rain and underground water have gradually molded the soft limestone here into a series of weird-looking peaks and fantastic grottos. We took a boat ride on the Li River to view some of these most unique landscapes in China, including the Seven Star and Reed Flute caves.

The next stop on our trip was Hangzhou, a scenic city situated on the beautiful West Lake. A boat ride on this lake is a must for any visitor. Many springs are located in this area, including the Tiger Spring, Jade Fountain, and Dragon Well, where the exotic Dragon Well tea plantation is located. Another attraction is the Hall of Heavenly Kings, in the Ling Ying Monastery, which contains a gilded camphor-wood Buddha eighty-two feet in height. While in Hangzhou, we stayed in a Guest House along with members of the Canadian Parliament and also were invited to attend an evening program with them.

In 1979, Shanghai was already a city of 11 million people. Professor Su Buchin, the President of Fudan University in Shanghai, was the person who had formally invited me to China. However, Professor Yang, Chairman of the Department of Nuclear Science, acted as my host while I visited Fudan University. His institute had an active experimental program centered around a 4.5-megavolt van de Graaff accelerator. At the time of my visit, Fudan University and the University of Beijing were considered the two leading universities in China. Fudan University officials urged me to give one of my three lectures on graduate education at the University of Rochester. (Professor Yang's daughter and only child, as well as Premier Deng Xiaoping's youngest son, Deng Zhifang, later came to the University of Rochester for their graduate studies.) There was at this time already considerable interest by the Chinese in graduate education in the West. The feeling prevailed that within a year or two, Chinese graduates would be able to compete in Western graduate schools. Educational levels had fallen to very low levels during the Cultural Revolution and were only beginning to recover at the time of my visit. The previous year a national examination had been instituted for entrance into a university. During the summer of 1979, twenty times as many people took the examination than there were places in the first-year class of all universities in China.

While in Shanghai we stayed in the Jing Jiang Hotel and had ample opportunity to visit many interesting sections of the city. Nanzing Road, the main shopping area, has a population density that is unbelievably high. We also enjoyed our visits to the old section of Shanghai and to the Shanghai Industrial Exhibition Hall. Dolly had the experience of viewing two major operations where acupuncture was administered as the only sedative. She was in a small glass-enclosed room, used as a viewing area, and located in an elevated position with a close and direct view of the operating table. One evening, Professor and Mrs. Yang entertained us in their home. Their apartment was very spacious by Chinese standards, however, rather sparsely furnished.

In Shanghai, I also visited the Institute of Nuclear Research. This institute had a cyclotron and was, at the time of my visit, in the process of constructing a new tandem van de Graaff. This laboratory has some 900 employees and is financed by the Academia Sinica. This is in contrast to the universities, which report to the Ministry of Education. This sharp division of research support between two agencies resulted from advice received from the USSR, where such a division of support is well-known in Soviet countries. Although the Chinese were beginning to realize the deficiencies of this system, they as yet had not changed it at the time of my visit.

Our next visit was to Lanzhou in Western China and it had to be planned in a timely way. The jet (Trident) flight from Shanghai to Lanzhou went only once a week on Friday morning. In Lanzhou, I lectured both at Lanzhou University, where Professor Wong Ying-Chang was director of the Department of Modern Physics, and the Institute of Modern Physics, where Yang Cheng-Chung was the director of the laboratory financed by the Academia Sinica. This latter institute was the most important laboratory in China for the study of heavy-ion nuclear science. They had modified a cyclotron (originally from Russia) for heavy ions and were using this machine as an injector into a new four-sector cyclotron. Some modern research was being carried out here and considerable development work on accelerator design. The university was geographically near the Institute of Modern Physics; however, it had far less advanced facilities. Lanzhou is an important industrial center of 2 million people in the Northwestern province of Kansu (or Gansu). The landscape reminds me of New Mexico, and the city spreads out for miles along the Yellow River. Much chemical industry is here, and the pollution is extremely bad due to the city lying between mountain ranges with little turnover of the air. Most people wore breathing masks since the quality of the air was so bad that it limited severely one's sight. This area was suffering also from a long drought at the time of our visit and food was in short supply (it was said that 20 percent of the population was on the brink of starvation).

We took advantage of our Western location to make a stop at Xi'an (Sian) on our return trip east to Beijing. Xi'an is an historic city and an important cradle of Chinese civilization. It is one of the more famous ancient capitals of China and had a population of 2.5 million people at the time of our visit. One of China's significant archaeological finds of the 1970s was the thousands of life-size terra-cotta warriors and horses guarding the main entrance of the tomb of Chin Shihhuang Ti, the first emperor of a unified China and builder of the Great Wall (Ch'in Dynasty 221–206 BC) The horses and warriors were discovered by chance in 1974 (five years before our visit) by peasants digging an irrigation well. Officials estimate that 6,400 figures are buried here. A huge new hangar-like hall the size of two football fields covered this area, which was still under excavation at the time of our visit. This museum, about to open in 1979, houses the greatest find of the century (it is reported that 700,000 men built this massive tomb). Other important highlights in Xi'an are the Bell Tower, Big Wild Goose Pagoda, Sian Hot Springs, and the Pan P'O village, a site of culture in 6,000 BC. Seeing these marvelous terra-cotta figures in Xi'an so soon after their discovery was probably the highlight of our trip.

Our next and last stop was Beijing. Here, I gave one lecture at the university and two at the Institute of Atomic Energy. All of my lectures lasted for three hours with a ten-minute intermission. In 1979, the scientists in China were aware of the scientific literature in the West but lacked modern facilities to compete on forefront research in nuclear science.

The University of Beijing, one of the best in China, has a beautiful campus and in 1979 had twenty-two departments (twelve in the natural sciences, seven in liberal arts, and three in foreign languages). At the time of my visit there were about 8,000 students (140 foreign), and approximately 600 postgraduates, however, they had plans for rapid expansion. The department of Professor Wu Ji-Ming, my host, had been moved to western China during the Cultural Revolution and had only returned to Beijing during the spring of the year of my visit; this group was just beginning to move back into modern research. Most of the students at Beijing University at that time lived on campus, six to a room.

The Institute of Atomic Energy (Academia Sinica) outside Beijing was, in 1979, one of the best Institutes in China. It was headed by Professor Wang Gon Tsang. He had spent a long period in Dubna (USSR) and returned to China in the mid-1960s to direct the Chinese atomic bomb projects. Wang had become director of this institute only two or three years before my visit, and had recently headed a delegation of Chinese scientists on a visit to the United States. During my visit to this institute, I was given a summary sheet containing the dates and types of nuclear detonations of the twenty nuclear tests conducted by China up to March 15, 1978. China exploded its first atom bomb on October 16, 1964, and its first hydrogen bomb on June 17, 1967. On October 27, 1966, China conducted a guided nuclear missile weapon test. The September, 2008, issue of *Physics Today* has a map on page 49 showing locations of the early nuclear facilities and test sites in China. This article also describes a visit Professor Yang, my host in China, made to Los Alamos in June 1988.

Beijing is the most advanced of Chinese cities. Some of the interesting sites we visited included The Great Wall, which transverses some 4,000 miles of mountains and canyons. The construction started in the Chou Dynasty (1129–249 BC), work continued in the Ch'in (221–209 BC), Han (206 BC–AD 270) and Tang (AD 618–905) dynasties. The Wall is some 22 feet high, 21 feet wide at the base, and 18 feet wide at the top, enough for five to six horses to run abreast. We also enjoyed seeing the Ming Tombs of thirteen emperors of the Ming Dynasty (AD 1368–1644), Summer Palace, Temple of Heaven, and the Imperial Palace (Forbidden City).

Our schedule was planned so that we would be in Beijing for the thirtieth anniversary of the inception of the People's Republic of China, proclaimed by Mao Zedong on October 1, 1949. On September 30, 1979, we were fortunate to be present in Beijing for the elaborate celebrations. We were invited to the national banquet in the Great Hall where 5,000 people gathered for this gigantic event. Various contests had taken place in previous weeks all over China and the winners of these competitions from the different provinces were invited to Beijing to participate in these national celebrations. A long table at the head of the Great Hall was occupied by the more influential Chinese political dignitaries. The more important guests at the banquet were assigned seats at tables near the head table. Guests of progressively lesser rank were assigned tables, which were proportionally farther and farther removed from the dignitaries' table. We were assigned a table along with the President of Beijing University and his wife quite near the head table, while the American Ambassador to China was assigned a distant table. It was explained to me that membership in the National Academy of Sciences trumped an Ambassadorship. Chairman Hua Guofeng spoke at the banquet.

Mao launched his Cultural Revolution in 1966, and Deng Xiaoping was sent to rural Jiangui province to work in a tractor factory. Red guards and leaders of the Cultural Revolution grasped power and drove China into great confusion for some ten years. In September, 1976, Mao died and the so-called Gang of Four were discredited. Jiang Qing, Mao's widow, was the leader of the Gang of Four. Deng Xiaoping had recovered power and was *Time's* 1978 Man of the Year. The year 1979 was an interesting time to be in China. It was a time when the Democracy Wall was popular. We continually heard, in all kinds of contexts, the denunciation of the Gang of Four. "Teach yourself" texts on physics, chemistry, and mathematics were best sellers among 1,500 books on popular science. During thirtieth-anniversary speeches and news articles the emphasis was to emphatically convince the people that "Mao Tse-tung thought is not final and eternal truth." The commentary said, "Three years have already passed since the disgrace of the Gang of Four" radicals but "a certain number of cadres still have fixed ideas." "Science and myths are incompatible." "We must adopt a new attitude to Mao Tse-tung." Mao Tse-tung's thought was not a "sublime truth" or something "unalterable and fixed."

Tired from our rigorous schedule in China, however with lasting memories of the Great Wall, the terra-cotta sculptures in Xi'an, and the festive events in the Great Hall, we returned to Hong Kong for a couple of days of rest before flying back to Rochester.

Offers for positions at other institutions continued to arrive. On the administrative side an offer of a departmental chairmanship at the University of Washington, Seattle, came in December, 1974; a vice presidency from SUNY Albany, in July of 1978; and during the 1974–75 period I was recruited by the Argonne National Laboratory to return as Associate Laboratory Director for the Physical and Mathematical Sciences. The Argonne offer was difficult to turn down as I held fond memories of my employment there from 1949–67. However, the offer I found most difficult to decline came from Michigan State University (MSU). I mentioned previously that they had tried to persuade me to spend the summer of 1970 there. In December of 1977, they made another attempt to entice me to spend time at MSU by inviting me to be a Visiting Professor for the year of July, 1978, through June, 1979. In May, 1978, a letter arrived stating that MSU had established a new John A. Hannah Professorship in the area of heavy-ion science (John A. Hannah was a former president of MSU), the first in the physical sciences. The letter went on to state that this was the second such professorship at MSU and that "these are the most distinguished faculty positions at the University. . . . Duties associated with the Hannah chairs are not pre-formulated, the concept being that duties will be adjusted to match the desires of particular applicants in order to have an arrangement which will maximize the scientific effectiveness of the person in question."

On July 28, 1978, the formal offer letter from MSU was sent to me. In addition to salary, the offer included a sizeable annual allocation of funds from the income from the John A. Hannah Professor endowment that could be used in support of research activities. MSU was in the process of developing a world-class nuclear research facility by coupling two superconducting cyclotrons. We considered the offer very seriously, making three trips to MSU, and went as far as to look at vacant lots and discuss building plans with contractors.

During the time period of 1974–83, I was heavily engaged with committee assignments both inside and outside the university. In addition to the normal load of college and departmental committees, I will mention two examples of my more time-consuming committee assignments. In 1977, I was appointed to serve on the Curriculum Review Committee of the College of Arts and Sciences. Members of the committee shared "concerns regarding the quality and appropriateness of the educational experience which we currently offer our undergraduates. These concerns range from modes of more effectively imparting to our students verbal and analytic skills, to proper preparation of our students for leadership roles in society." Plainly put, the charge to our committee "is to prepare a program for strengthening those areas of the curriculum which lie outside of the concentrations—a program which can be implemented within the resources of the College, and which is designed to enlist on its behalf the enthusiastic support of the Faculty, while expanding in our students a sense of being well served."

On October 30, 1981, I was informed by President Sproull that the "Faculty Senate is organizing a new *ad hoc* committee on Graduate Education—and at a Senate Steering Committee meeting you were chosen as a member." The charge to the committee will be as follows: "The Committee will be asked to evaluate the status of and trends in PhD studies in the various colleges, and to review existing plans for future development in particular taking into account likely changes in outside funding. The Committee will also be asked to report its findings to the Senate with any recommendations it desires to add." Our detailed and lengthy report was submitted to the Faculty Senate on April 19, 1983. I'll give an impression of the report by quoting from the introduction. "This Report of the Ad Hoc Committee on Graduate Education to the Faculty Senate of the University of Rochester is in two parts. In Part I there is a general review of the developments of the period from 1970 to the present. This review at some time considers most parts of the University, but its focus is on the College of Arts and Science where about 60 percent of the doctoral Candidates are normally present. . . . This review considers the grounds for supporting graduate students as well as history of graduate recruitment and graduate support over the recent period. In addition, it examines the national reputation and the remuneration of the faculty, and, in a very summary manner, some facts about the financial resources of the University for undertaking its teaching and research responsibilities. . . . Throughout Part I in order to measure the performance of Rochester programs we make comparisons to other universities. Most of the comparisons are made with two universities . . . Princeton and Northwestern. . . . We have also made some comparisons with a group of six universities which share many characteristics with Rochester." Part I of our report "concludes with a number of specific recommendations which are intended to suggest ways in which the graduate program at the doctoral level might be improved."

"Part II of the Report is made up of four summaries of developments in particular areas of study." These four summaries are labeled, "Graduate programs in the Social Sciences, Graduate Education in the Humanities, Graduate programs in the Physical Sciences and Graduate education in the Biological Sciences." Individual members of the Committee wrote each of these summaries, and I wrote the summary on physical sciences. The entire report can be found in my Correspondence Volume, 1978–83.

My committee assignments for professional activities outside the university were numerous during these years. In 1974, I was appointed to chair the National Academy of Sciences Committee on Nuclear Science. This committee had some fifteen members from academia, national laboratories, and industry. Charles K. Reed, an employee of the National Research Council of the National Academy of Sciences, served as Executive Secretary of our Committee. The charge to our

Committee on Nuclear Science is very long, and I quote only the first sentence, "The Committee on Nuclear Science seeks to maintain, on a continuing basis, an assessment of the status, progress, new directions, opportunities and goals of nuclear science."

The work of the NAS Committee on Nuclear Science was carried out through ad hoc panels appointed by the committee. In addition to committee members, these panels contained members from the entire nuclear science community, selected for their special expertise. The committee worked closely with the Division of Nuclear Physics of the American Physical Society, the Division of Nuclear Chemistry of the American Chemical Society, and other concerned professional societies as representatives of the nuclear community. Our NAS Committee on Nuclear Science met monthly for two days in a conference room at the NAS building to hear and study panel, committee, and member reports; hear testimony; and plan new studies.

In 1978, the Department of Energy (DOE) and the National Science Foundation (NSF) established a joint DOE/NSF Nuclear Science Advisory Committee (NUSAC). I was appointed as one of the charter members of NUSAC. The appointment letter stated, "Your appointment to this Committee is as an individual, and it is your personal role which is solicited, rather than representation of your institution or of any component of the nuclear science community." I was immediately notified that I was also appointed to the Facilities Subcommittee of NUSAC. A short time later, I was appointed to chair the subcommittee on University Research and Education in Nuclear Science. In October, 1982, the subcommittee issued a comprehensive 46-page report, which included a set of recommendations to strengthen nuclear science education and research. These assignments illustrate how much of the work of NUSAC was accomplished through these subcommittees. However, NUSAC as a whole met frequently and each summer sponsored a week- or two-week-long workshop in which the current status of nuclear science was evaluated and recommendations for the future health of the field were developed in a comprehensive report.

Having been employed for many years at a large national laboratory, I was frequently appointed to serve on advisory and review committees at national laboratories. These appointments included being a member of (1) the Review Committee for the Physics Division of the Argonne National Laboratory, 1973–76 (Chair in 1975); (2) the Lawrence Berkeley Laboratory Nuclear Science Division's Visiting Committee, 1976–78 (Chair in 1977 and 1978); (3) the Los Alamos Meson Physics Facility (LAMPF) Program Advisory Committee, 1976–79; (4) the Argonne Universities Association Review Committee for the Chemistry Division, 1979–82 (Chair in 1981); and (5) the Scientific Review Committee of Nuclear Chemistry Division of the Lawrence Livermore National Laboratory, 1978–83.

In 1981, I was appointed to the Scientific and Educational Advisory Committee (SEAC) for the Lawrence Berkeley Laboratory. This committee was made up of an equal number of members from inside and outside the California Research University system. The responsibility of SEAC was "to advise the President and Director of the laboratory on performance, program, and plans of the laboratory, to advise the Director and the Berkeley Chancellor on the relationship between the laboratory and the Campus in scientific and educational matters, and to make periodic reports to the President, based on continuing reviews." SEAC reported directly to and met for an extended time, following each of our meetings with President Saxon, who was the chief administrator for all the California Research Universities.

During this period I served also on two university visiting committees in Texas, the User Committee of the National Superconductivity Laboratory 1980–82 (Chair for 1981 and 1982), the program committees of the Nuclear Physics Section of the American Physical Society, and on the Editorial Board of the *Physical Review C*.

In June of 1974, while still on sabbatical in Munich, I learned that I was the recipient of the 1975 American Chemical Society's Award for Nuclear Applications in Chemistry (now called the Glenn T. Seaborg Award for Nuclear Chemistry). The award was presented on April 7, 1975, at the Philadelphia ACS meeting. The citation in part read "for his research on many aspects of the nuclear fission process, in particular the properties of excited states at the fission barrier as well as the competition between fission and other modes of decay of highly excited nuclei, and for his contributions to the determination of nuclear level densities as a function of excitation energy." One of the pleasant rewards from the publicity of such an award is the number of letters it generates from friends and acquaintances from the distant past. During Calvin College's Commencement exercises on May 24, 1975, held in Grand Rapids, Michigan, I was presented the Distinguished Alumni Award. This award "is presented to alumni who have made very significant contributions in their professions or field of service and are recognized for such accomplishments by their associates." My college roommate, James Bere, who became President of Borg-Warner, had won the award a few years previously. I joined the American Association for the Advancement of Science (AAAS) in late 1980. On January 16, 1981, I was notified that at its "meeting on January 7, 1981, the AAAS Council elected you a Fellow of the Association. . . . A Fellow of the AAAS is defined as a member whose efforts on behalf of the advancement of science or its applications are scientifically or socially distinguished." I had previously received Fellow status in the American Physical Society in the 1960s.

In April, 1976, I was elected to the National Academy of Sciences (NAS). I first learned of my election, minutes after my election, by a phone call to my room in a Washington hotel while I was attending the American Physical Society's Annual spring meeting. I was about to leave my room for a scientific session as the phone rang shortly after 8 AM. On the phone were a group of academy members offering their congratulations, including Jake Bigeleisen,

Glenn T. Seaborg, Fred T. Wall (my thesis advisor at the University of Illinois), Tony Turkevich, and others. The formal ceremonies of induction into the NAS and the signing of the NAS registry occurred the following April, 1977. As a member working in an interdisciplinary area of research, I had the option at that time to join either the chemistry or physics section of the academy, having been invited to join both sections. I initially joined the chemistry section. In recent times the policy of being a member of only one section has changed, and I'm currently also an affiliate of the physics section. The election process is confidential; however, I became aware that I was being considered as an intersectional nominee when John A. Wheeler, a distinguished physicist and co-author with Niels Bohr of the first theoretical paper on nuclear fission, sent a copy of his supporting letter to me, presumably through some kind of secretarial mistake. The NAS meets annually in Washington, DC, during the last week of April, a delightful time of the year to visit Washington. The meeting occurs at the academy's elegant facility at 2101 Constitution Avenue N.W. (they have a second meeting facility at 500 Fifth Street N.W.). During the annual meeting, catered meals are served in a large tent placed on the academy grounds. Meetings begin late on a Sunday afternoon with cocktails and a fantastic spread of food, followed by a first-class concert. I have attended each of these annual meetings over the last thirty or so years.

X. The Years as Department Chair (1983–1988)

For many years I successfully declined the college Dean's proposals to become department chair. I had always assumed, however, that at some later and more appropriate time, I would fulfill my departmental responsibility by serving as chair. In 1983, I decided the time had arrived. A February 23, 1983, letter from Dean Hunter to the chemistry faculty announced that I had agreed to become chairman. This letter stated, "Your departmental deliberations were so swift and so efficient that I did not even have the opportunity to invite you to come individually if you wished to see me to talk over not only the chairmanship. . . ." On accepting the job, I decided to make a strong effort to improve the department. My goals were to strengthen physical chemistry, actively nominate faculty for appropriate awards, increase graduate student stipends and enrollment, renovate and expand considerably the research space in Hutchison Hall, and modernize and increase the departmental equipment and instrumentation. The latter goal required also a sizeable expansion of the departmental service and instrumentation space, as well as the addition of new full-time technical staff, so necessary in maintaining the departments' array of new instruments.

A. My Administrative Venture

On becoming chair, I appointed a full-time administrator of chemistry (the title was later changed to assistant chair) with the dual responsibility of handling the more routine day-to-day administrative functions, associated with a smooth and efficient departmental operation, and to supervise the department's non-academic staff, which eventually in my term increased to twenty full-time employees. In addition, a full-time administrator was appointed to work with a faculty committee on graduate student recruitment and supervision. The departmental secretary also plays a key role by effectively regulating the chair's daily schedule of meetings and other activities. Another major function of this secretary was to screen the daily departmental correspondence and assist in preparing the department's response. With the above administrative structure in place, I was able as chair to focus on major issues and challenges associated with the improvement of the department's status.

After a short introductory period in which the department's administrative structure was put into place and tested, I worked out a schedule where on most days I went to my research office by mid-afternoon. Udo Schröder performed double duty during this period in supervising all the nuclear students during my absence. Time was at a premium during this five-year period and I had to ration it wisely.

The degree to which I fulfilled my goals as chair needs to be answered by others. I will attempt, however, to summarize some of what I consider to be my major accomplishments as chair. First on the negative side, I failed in my attempt to recruit a renowned physical chemist from the Netherlands. Even so, by this experience, I was emboldened and greatly encouraged by the college's support in my development of a very lucrative startup package, probably one of the best in the college's history. On the positive side, I assisted in launching the careers of three outstanding physical chemists by supplying them initially with the necessary monies for equipment and instrumentation and renovating ample laboratory space to their specifications. During my term as chair, I also renovated laboratory space for new junior and senior organic chemists as well as provided their startup funds. I was able also to obtain college funds for the development of a considerable amount of additional new laboratory space for successful faculty groups needing additional research space. During my term as chair, the supervision of renovation projects caused me the most frustration. Not being allowed to seek outside competitive bids for projects, I had to rely on university personnel for all renovations. This led not only to inefficiencies and construction delays but also exorbitant costs for the university.

Another department endeavor occurring during my tenure as chair, and requiring the commitment of a considerable amount of new laboratory space was a joint proposal with scientists from Eastman Kodak and Xerox for a Science and Technology Center for Photoinduced Charge Transfer, to be funded by the National Science Foundation. The development of this proposal was time consuming, requiring numerous meetings with the executives of both companies. One of the major strengths of our proposal was its clear demonstration of a truly interactive academic–industrial collaboration. Our proposal was in competition with 322 other proposals for National Science Foundation Centers. Of these, some 45 proposals were selected for a formal site visit representing the second phase of the merit review process. Our site visit occurred on June 16 and 17, 1988. Our proposal was among the one-fifth of those having a site visit that eventually were selected for funding. David Whitten played an important role in developing the scientific program for the center, and he became the first director.

During the 1983–88 period, five new assistant professors joined the department. Two faculty were promoted to associate professor with unlimited tenure and six faculty were promoted to professors. It is important to point out that the average age of the chemistry faculty in 1988 was only 44 (also the median age). Thus the faculty was in a strong position for further growth in excellence over the next decade. A number of awards and honors were received by faculty during my term as chair, including (1) the American Chemical Society's Award in Pure Chemistry (one of the most prestigious Awards of the ACS); (2) an NSF Presidential Young Investigator Award (this is a five-year award with a possible total support of $100,000 per year); (3) two Camille and Henry Dreyfus Teacher-Scholar Awards; (4) two Camille and Henry Dreyfus Distinguished New Faculty Awards; (5) an Edward Peck Curtis Teaching Award; (6) a John S. Guggenheim Fellowship; (7) three Alfred P. Sloan Foundation Awards; (8) a Fulbright Fellowship; (9) three American Physical Society Fellowships; (10) a NIH Senior Postdoctoral Fellowship; (11) the ACS Rochester Award; (12) the Exxon Education Foundation Award; (13) a Fellowship from the Royal Society of London; (14) a Humboldt Award; and (15) a University Mentor Award.

We decided early on during my years as chair that the highly competitive nature of graduate student stipends necessitated that we find some funding for our students outside the traditional university sources. By 1985, stipends had been raised substantially to $9,000 ($11,000 for the better Fellowship students). The 1985 class consisted of thirty-four students with an average grade point average of 3.4 (on a scale of zero to 4.0), and with median graduate record examination scores of 690 (quantitative) and 610 (verbal). These students came from a number of major universities, including Arizona, Michigan, Princeton, Brown, Emory, Cornell, SUNY Buffalo, Rutgers, North Carolina, and Pennsylvania. It was one of the largest and most outstanding entering classes in the department's history. In 1986, we again recruited thirty-four students. This was a major accomplishment, given the nationwide decline of the applicant pool. During the last three years of my term, we had a steady-state number of 130 graduate students, compared with some 70 a decade earlier.

New departmental equipment and instrumentation added during my five-year term as chair included a 500-MHz and two 300-MHz nuclear magnetic resonance (NMR) spectrometers, a GC mass spectrometer, X-ray diffractometer, two VAX computers (one for scientific computing and one for upgrading the administrative record-keeping), a single photon counting system, a scanning tunneling microscope, a Fourier transform infrared spectrometer, a neutron meter, and a large amount of laser equipment. Funds for these instruments, totaling between two and three million dollars, were raised from the National Institute of Health, the National Science Foundation, the Department of Energy, the Department of Defense, the University of Rochester, and industrial and alumni donations.

The utilization of most of this new equipment and instrumentation required our raising substantial additional funds for renovation of laboratory space to house it and the hiring of new technical staff to maintain it. The departmental laser equipment was housed in dedicated space on the first floor of Hutchison Hall and the laser facilities were placed under the management of a technical staff person with an advanced degree in optics. The 500-MHz NMR spectrometer and associated computer equipment were installed in a full laboratory module in the basement of Hutchison Hall. The two 300-MHz NMR spectrometers were installed on the fourth floor of Hutchison Hall in a several-module-dedicated space for department instruments. Highly qualified technical staff were hired to supervise the use of and maintenance of these and other instruments. At this point our technical staff had increased to six full-time employees.

Early in O'Brien's term as president, he produced a document entitled "University of Rochester Five-Year Objectives." I was given an opportunity to comment on an early draft of this report and did so promptly. Here, I report on some of my comments on the role of technical staff in a research university. I wrote to O'Brien, "One important group is missing from point 8. Highly trained and skilled technical specialists play an important role in modern research departments. The University of Rochester's present top grade and salary for these people must be substantially improved. We recently lost a talented NMR-mass spectroscopy-computer specialist to another university at a salary some $15,000 above our top salary!" By this action I played a role in upgrading the university's pay grades and salary structure for technical personnel in all research departments.

Two new chemistry sequences were initiated in the fall of 1987. The first-year chemistry offerings consisted of Chemistry 103 and 105 in the fall semester, and Chemistry 104 and 106 in the spring. These two courses in each semester follow the same basic curriculum, differing only in the depth of coverage and mathematical skills used in developing and applying concepts to specific problems. Chemistry 105 and 106 were enriched versions of Chemistry 103 and 104, respectively. The program in the first year offered complete flexibility and avoided the problem of having to choose a specific sequence before a student's interests have developed adequately. A similar two-track system was developed for the second year of Organic Chemistry. One of our alumni made a substantial endowed donation to the department to provide for an annual revitalization of the undergraduate laboratories.

During my final year in office, the Chemistry Department research budget had increased to 4.35 million dollars, by far the largest such budget in the department's history. In terms of the amount of federal research funds per faculty member, we ranked near the top of all chemistry departments in US universities. I left office with some satisfaction that the department had improved considerably and was well positioned with a young and talented faculty with enormous potential for future advancement.

On June 30, 1988, I sent my last annual letter as Chair of the Department to alumni and friends of the department. I used this occasion to review briefly some of the activities that occurred during the five-year period (1983–88) in which I served as Chair of the Department. On July 1, 1988, the department celebrated the completion of my term as Chair of Chemistry with a festive dinner in the Gallery Café located in the Memorial Art Gallery.

B. Damped Nuclear Reactions

The ^{144}Sm + ^{84}Kr reaction was studied (225) at laboratory bombarding energies of 470, 595, and 720 MeV, corresponding to energies 1.2, 1.6, and 1.9 times the entrance-channel Coulomb barrier, respectively. Correlations between angle, kinetic-energy loss, and charge of the projectile-like fragments are investigated. A probability for orbiting is observed at all bombarding energies, with fragment kinetic energies indicative of large exit-channel deformations. This study shows that energy dissipation (up to about 100 MeV) and charge exchange are correlated in a manner which appears to be independent of initial relative ion velocity. The variances in the fragment charges as a function of energy loss are compared with a classical dynamical transport model.

As part of a *Treatise on Heavy-Ion Science,* edited by D. Allan Bromley, Schröder and I published in 1984 a 620-page comprehensive review article (228) entitled "Damped Nuclear Reactions." Quoting from a paragraph in the Introduction we stated: "The growing scientific interest in damped nuclear reactions is attracted by the many opportunities they provide for a study of phenomena occurring in nuclear matter under extreme conditions with respect to shape, intrinsic excitation, spin, mass-to-charge, density, etc., and the possibility to control these conditions by selecting appropriate initial and final reaction parameters. Particular goals are to discover the simple modes of nuclear excitation produced in damped nuclear collisions, their foundation on the microscopic processes, and their eventual decay, i.e., the cooperative phenomena and relaxation processes occurring within a small quantal system that is initially far from equilibrium. It is inescapable to notice the analogies of these problems with those addressed by other branches of science such as statistical and chemical physics, which invites an interdisciplinary approach to a profound understanding of damped-reaction mechanisms. Damped nuclear reactions, however, not only provide the means to study modes of nuclear motion inaccessible by other methods—they are also amenable to the production of exotic new nuclides far off the beta-stability line." The more than 1,000 references cited in this manuscript are confirmational evidence of the intensive experimental and theoretical activity in this new field of study over the last few years.

On April 17, 1984, I received an invitation to speak at the centenary of Niels Bohr. The letter stated that "it is important for us at this occasion that the meeting should take place at the Niels Bohr Institute. This means that we can have at the most 120 participants . . . participation will be by invitation only." I was not only delighted to be invited to this historic event of the 100th

anniversary of Niels Bohr's birth, but especially honored to present one of the lectures. I chose to discuss Nuclear Transport Phenomena in Low-Energy Heavy-Ion Collisions (229).

A paper describing the mass, charge, and energy transfer in damped reactions was presented at a Rochester conference (235). The abstract stated: "Certain features of transport phenomena in damped reactions are reviewed, with emphasis on the dynamic and static forces driving mass and charge exchange. Important ingredients of a one-body reaction model are discussed, in which transport is induced by the exchange of individual nucleons. The model, incorporating window and (nominal) wall friction, successfully describes many reaction aspects for a number of systems but is challenged in the reproduction of drifts in mass and charge asymmetry exhibited by systems featuring relatively large static driving forces."

A ^{58}Ni target was bombarded with ^{165}Ho ions accelerated to energies of 970 and 1075 MeV (237). Coincident reaction fragments were detected with two position-sensitive multiwire avalanche detectors (PSADs). The position was determined via the delay-line readout of the wire plane. This position information from both PSADs was used to determine the angles θ and φ of each of the correlated fragments, as well as their flight distances from the target to the detection site. The detectors produced timing signals as well, providing information on the time of flight difference of the correlated fragments. The determination of the two angles (θ_1 and θ_2), of the two corresponding flight distances, and the difference in times of flight, constitute a kinematically complete measurement of two-body reaction events. Correlations between energy, mass, and angle show that a small fraction of the events have fully relaxed kinetic energies, masses intermediate between those of the projectile (target) and symmetric fragmentation, and angular distributions with a marked forward-backward asymmetry.

Experimental neutron multiplicities and spectral slope parameters are compared with evaporation calculations for the damped ^{165}Ho + ^{56}Fe reaction as a function of E_{loss} for different degrees of energy relaxation of the dinuclear system (238). Detailed comparisons for two different primary reaction fragment distributions show that energy partition of the primary fragments is energy-loss dependent and, for small energy losses, far from thermal equilibrium.

The redistribution of kinetic energy is studied (240) for the damped reactions ^{139}La + ^{40}Ar at a bombarding energy of 10 MeV/nucleon. Energy spectra and multiplicities of neutrons measured for this reaction are indicative of temperature disparities between the two primary damped reaction fragments. In addition preequilibrium emission of neutrons is observed in this reaction with a multiplicity depending on the degree of energy damping.

A large neutron multiplicity meter has been used in a study of energy partition in the reactions ^{165}Ho + ^{56}Fe and ^{209}Bi + ^{56}Fe at bombarding energies of 7.0 to 8.75 MeV/nucleon (245, 240). The neutron multiplicity is measured employing a highly efficient, near-spherical neutron multiplicity meter

(NMM) of one meter diameter (see figure 43). The NMM consists of a tank filled with 220 gallons of NE-224 liquid scintillator, viewed by eight photomultipliers. Neutrons entering the sensitive volume of the NMM are thermalized by multiple scattering processes within the liquid and finally are captured by a gadolium component of the scintillator. Light produced by the neutron capture gamma rays in the scintillator is detected by the photomultipliers. The statistical slowing-down process and the subsequent neutron diffusion process lead to a spread over several tens of microseconds of the capture times of the individual neutrons of a multi-neutron event. The number of neutrons can then be measured simply by counting the number of delayed light pulses following a reaction event.

Since the available neutron multiplicity meter (NMM) is not directionally sensitive, the NMM is operated in two different geometries with respect to the target-detector assembly. In the "closed" geometry, the NMM measures the total multiplicity $m = m_L + m_H$ of neutrons from light (L) projectile-like fragment and the heavy (H) target-like fragment, respectively. The lower portion of this figure exhibits the NMM in an open geometry, aligned with a solid-state detector telescope measuring the projectile-like fragment. Here, the NMM measures with a relatively high probability the well-focused neutrons from the projectile-like fragment.

Both the average energy partition and the magnitude of its fluctuations for the above two reactions are found to be uncharacteristic of thermal equilibrium. The experimental data are compared to the predictions of a transfer-induced dinuclear heat convection mechanism, and dinuclear thermal relaxation times are found to range between 10^{-22} to several times 10^{-21} seconds.

Projectile-like fragments were studied with high atomic number Z and atomic mass A resolution for the four damped reactions $^{238}U + {}^{40}Ca$, ^{48}Ca, ^{58}Ni, and ^{64}Ni at 8.5 MeV/nucleon bombarding energy (242, 243). These dinuclear systems exhibit widely different driving forces. Qualitative agreement between experiment and dynamical nucleon-exchange transport theory is obtained for the first and second moments of the fragment atomic (Z) and neutron (N) number distributions. However, systematic differences exist between the experimental and theoretical average values of Z and N suggesting the need for more realistic driving forces.

The charge equilibration process in damped reactions has been examined by measuring discrete nuclidic distributions for the $^{238}U + {}^{58}Ni$ and ^{64}Ni reactions at a bombarding energy of 8.5 MeV/nucleon (244). The data demonstrate that in these very asymmetric systems the evolution of the nucleon-exchange process as a function of energy loss depends strongly on the N/Z value of the projectile and the corresponding gradient in the potential-energy surface. Comparison of the data with transport model calculations shows qualitative agreement with the N and Z centroids, variances, and correlation coefficients. Absolute discrepancies exist, however, which suggest the need for improvement in the model.

EXPERIMENTAL SETUP

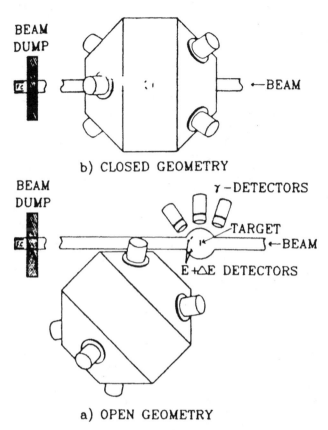

b) CLOSED GEOMETRY

a) OPEN GEOMETRY

Figure 43. Schematic illustration of our first neutron multiplicity meter (245).

Some evidence has been obtained for the emission of alpha particles from the neck region during the final stages of damped reaction processes with low energy losses (234), similar to such emission during the fission process.

C. Fission-Like Reactions

The measurement of the fraction of the projectile momentum transferred to the composite system in a fusion-like reaction can be made by measurement of the correlation angle between the two fission fragments. Such correlation angle measurements were made of fission fragments produced by ^{20}Ne-induced fission of ^{165}Ho, ^{181}Ta, ^{197}Au, ^{209}Bi, and ^{238}U at a projectile bombarding energy of 290

MeV (221, 230, 233). One of the aims of this work was to determine the cross section of events associated with complete linear momentum transfer, which is assumed to be the fusion cross section. A schematic diagram of the scattering chamber, including the ionization counter, recoil detector, and beam monitoring detectors is shown in figure 44A.

The in-plane correlation angle spectrum for coincident fission fragments for the ^{181}Ta target is displayed in figure 44B. The five arrows on the top of the solid curve correspond (from the left) to 100, 80, 60, 40, and 20 percent linear momentum transfer, respectively, that is, they signify the escape of up to four alpha particles. The analysis of the correlation-angle spectra assumes that the measured out-of-plane dispersion in the fission fragment correlation angles is due to particle evaporation before, during, and/or after fission. The average value of the FWHM of the out-of-plane dispersion is 10.7 degrees for the five targets, a value consistent with the experimental errors. This out-of-plane angle φ was used to fit each of the measured in-plane correlation-angle spectra with a series of Gaussian peaks, with centroids constrained to lie at the correlation angles predicted for fission at the most probable total kinetic energy, at a series of values of linear momentum corresponding to an increasing number of uncaptured alpha particles. The intensities of these various capture processes were adjusted to give the best fit to the in-plane correlation angle data.

The probability of 100 percent [P(^{20}Ne)], 80 percent [P(^{16}O)], 60 percent [P(^{12}C)], and 40 percent [P(^{8}Be)] linear momentum transfer for fission of a ^{181}Ta target, as deduced from fits to the experimental in-plane-correlation spectrum (see figure 44B), are 0.58, 0.31, 0.09, and 0.02, respectively. The experimental in-plane correlation-angle spectra for all the other targets were analyzed also in a similar way.

Assuming a sharp cutoff model, the maximum value of the angular momentum ℓ_{FLMT} that fuses is calculated from the measured value of σ_{FLMT}, the total fusion cross section for events with full linear momentum transfer. This experimental value of the angular momentum ℓ_{FLMT} is compared for a number of heavy-ion reactions with the maximum value of the angular momentum that fuses, ℓ_{fus}^{max}, as calculated with a classical trajectory model based on the proximity nuclear potential and one-body nuclear dissipation (see Subsection IX.B.3). In the trajectory-model calculations, the angular momentum ℓ_{FLMT} corresponds to rolling collisions with a value of angular momentum equal to $(7/5)\,\ell_{POCKETS}$. In figure 44C the ratio R of the experimental maximum angular momentum for complete linear momentum fusion, ℓ_{FLMT}, and the theoretical maximum value of the angular momentum that fuses, ℓ_{fus}^{max}, is plotted as a function of the available energy over the barrier in MeV per nucleon $(E_{c.m.}-V_B)/\mu$. When the limiting angular momentum is plotted in terms of the ratio R, one observes that this ratio shows the same general trend for both relatively light and heavy targets when plotted as a function of the available energy over the barrier in MeV per nucleon. When $(E_{c.m.}-V_B)/\mu$ reaches approximately 4 MeV per nucleon,

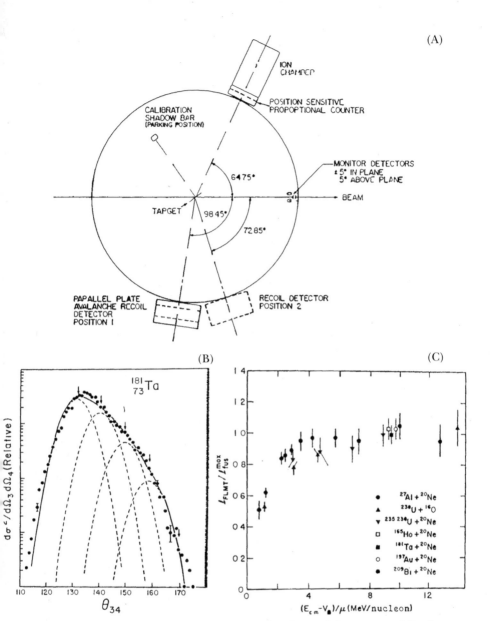

Figure 44. (A) Schematic drawing of scattering chamber and detectors used for the measurement of linear momentum transfer in fission-like reactions (230). (B) In-plane correlation angle data for 290-MeV ^{20}Ne induced fission of ^{181}Ta. The heavy line is the sum of Gaussian fits with FWHM equal to 10.7 degrees to each fusion component as shown by the dashed lines (230). (C) Systematics of the ratio $\ell_{\text{FLMT}}/\ell_{fus}^{\max}$ as a function of $(E_{\text{c.m.}}-V_B)/\mu$ in energy per nucleon. The angular momentum ℓ_{FLMT} is the limiting angular momentum corresponding to full linear momentum transfer. The quantity $\ell_{fus}^{\max}$ is the theoretical prediction of a one-dimensional classical dynamical model (223) for the maximum angular momentum that fuses (230).

ℓ_{FLMT} has reached its maximum value corresponding to an angular momentum near the rolling limit of a collision between the target and projectile based on the proximity nuclear potential and one-body nuclear friction. The present experimental values of ℓ_{FLMT} suggest that angular momentum dissipation in the entrance channel is an important factor in determining whether fusion with full linear momentum transfer will occur.

Fragment angular distributions are reported (218) for 220-MeV induced fission of ^{165}Ho, ^{197}Au, and ^{209}Bi. Theoretical expressions for fission fragment angular distributions at low and modest energies are discussed in Subsection IV.B.2. Comparison of the above high-energy heavy-ion-induced fission experimental results with the earlier described theory requires several important modifications. In the simple transition-state theory, as I_{sph}/I_{eff} approaches zero, K_o^2 approaches infinity and the fission fragment angular distribution becomes isotropic. In order to perform a realistic comparison of the high-energy heavy-ion experimental results with the rotating-liquid-drop model, one has to know the fissility parameter, nuclear temperature, and spin distribution of each of the sequence of nuclei undergoing fission following the initial formation of a composite system.

Incomplete linear momentum transfer (discussed earlier in this section) and emission of prefission particles in high-energy ^{20}Ne induced fission causes angular momentum de-alignment and large uncertainties in the spin distribution of the fissioning nuclei. An additional complication in heavy-ion-induced fission of high-Z targets at high energies is that the fission barrier has vanished for a large fraction of the angular momenta. When the fission barrier has vanished, one can no longer invoke the concepts of compound nucleus formation and unconditional saddle shapes. In this case the projectile and target amalgamate during the inward radial motion, and the composite system eventually leads to symmetric fragmentation. The stage at which the K distribution is frozen in for such systems may be near the turning point of the trajectory, where it spends a considerable fraction of its lifetime. Although such a postulate for the turning point in symmetric fragmentation playing the role of the saddle in compound-nucleus reactions is plausible, the K distribution may well be established at a later stage for these systems with large angular momenta. Furthermore, the presence or absence of a small fission barrier may be of little importance in determining where the K distribution is frozen in for rapidly rotating nuclei at high excitation energy.

After account is taken in the transition-state theory of both prefission particle emission and incomplete fusion (less than full linear momentum transfer), the rotating-liquid-drop model fails to account for the fragment angular distributions for heavy-ion induced fission. For such reactions, as ℓ_f exceeds I_{RLDM} $(B_f = 0)$ and the unconditional saddle-point energy has gone to zero, the saddle-point configuration plays less and less of a role in establishing K_o^2.

A statistical-scission model (SSM), first proposed by Ericson, is developed and applied to fragment angular distributions from heavy-ion induced fission, where the fission barrier has gone to zero (226, 231, 232, 236). Although the formal equations for fragment angular distributions in the statistical-scission and transition-state models are of the same structure, the variances in the distribution of angular-momentum projections on the fission direction are established at very different stages of the fission process in the two models. The SSM predicts angular distributions in reasonable agreement with those measured for heavy-ion-induced fission of some reaction systems where the fission barrier has vanished or is very small relative to the nuclear temperature. For a number of these systems, the transition-state model is inapplicable. The SSM, however, predicts the variance to have too weak an energy dependence, possibly indicating dynamical effects. Inclusion of pre- and post-scission particle emission and asymmetric fission does not completely remove this discrepancy. The large effective variances, deduced for systems with large spin and no fission barrier, are inconsistent with the concepts of "fast" or "preequilibrium" fission, occurring on a time scale too short for relaxation of the tilting mode. Some of the present discrepancies between the effective variances deduced from fragment angular distributions and those calculated with the SSM may be due to small contributions of incomplete fusion followed by fission or due to sequential fission.

I presented a paper entitled "Massive Heavy-Ion Reactions" at a symposium held on June 2–5, 1938, in Obninsk, USSR (241). This lecture describes selected topics associated with massive heavy-ion reactions. Included were discussions of energy dissipation and equilibration of fragments in damped reactions, angular distributions of fragments in fusion-fission reactions, and intermediate-type reactions with partial relaxation of the mass-asymmetry degree of freedom.

In the case of composite nuclei at moderate energies and angular momenta, such as those produced with light-ion projectiles, evaporation spectra are well explained in terms of standard statistical models employing optical model transmission coefficients in the description of particle evaporation. Over a period of time, many papers were published claiming that experimental charged particle evaporation spectra from heavy-ion fusion reactions, where higher excitation energies and angular momenta are involved, are no longer consistent with predictions of such models. Specifically, it has been suggested that in the latter cases the particle emission barriers are significantly lower than those expected from optical model transmission coefficients calculated for the respective inverse absorption channels. Contrary to these above claims, we showed (239) that the standard statistical model approach gives good agreement between calculated and experimental alpha-particle spectra from heavy-ion fusion reactions at high excitation energies and angular momenta.

The internuclear potential determined from heavy-ion fusion reactions at the fusion barrier distance was reported earlier (173). Others suggested later a technique for determining the internuclear potential at small separation distances inside the fusion barrier. In response, we pointed out the difficulties associated with the determination of the nuclear potential at distances inside the fusion barrier (227). We conclude that the critical distance model analysis of fission excitation functions at high bombarding energies does not give a model independent measurement of the internuclear potential at small separations of interacting heavy nuclei.

D. Other Activities

During my years as department chair, I served also on a number of influential university and government committees. First, I'll mention some of the time-consuming university committees. In early 1983, the University of Rochester's Board of Trustees announced the members of the Trustees' Committee for the selection of the university's new president and of the Faculty Advisory Committee to the trustee's committee. Robert L. Sproull had announced that he would retire from the president's post on June 30, 1984. The Faculty Advisory Committee consisted of one member from each of the Colleges of the University plus the Dean of the Medical School and the Dean of the College of Arts and Science. I was elected by the Colleges of Arts and Science to be its member.

The Faculty Advisory Committee worked very hard as evidenced by the fact that they screened some 260 candidates. Members of the Committee, usually in teams of two, made several trips in 1983 to interview candidates. In addition to conducting interviews, the committee held numerous meetings discussing the qualifications of various candidates. I retained the minutes of one of our meetings held on January 25, 1984. This document describes individual committee member's evaluations of the four candidates chosen for campus visits. Each committee member met with a different segment of the university to obtain feedback from faculty after the candidates' visits.

I developed a list of attributes expected for a leader of a complex research university and evaluated each of the four candidates chosen for visitation based on these criteria. Individual members of our Faculty Committee rated each of the four presidential candidates visiting the campus very differently. For example, some faculty members in their evaluations emphasized experience in a private institution. Others emphasized experience in a complex research university, independent of whether it be in a private or a public university. In my judgment only two of the candidates met the criteria I had developed. However, I was in the minority and the Faculty Advisory Committee recommended a candidate from a private college without any prior experience in a complex research university.

During this period I also served on a university committee with the charge to "examine the role and responsibilities of the University Dean of Graduate Studies and the make recommendations to the Council of Graduate Studies as to what that role should be in the future." Our committee issued a seven-page report on January 19, 1984. This report included an overall recommendation and three specific recommendations. The overall recommendation stated "the office of the University Dean for Graduate Studies should be strengthened so that the dean can act effectively as the principal academic spokesman for graduate education in the University. Consistent with this enhanced role, the graduate dean should have a trustees visiting committee for the graduate program so that the importance of graduate education can be stressed at the highest levels." The function of the graduate dean's office should involve (a) university policymaking, (b) control of a specific budget for university-wide recognition and support of excellence, and (c) administration of the PhD programs. Major new specific recommendations were made in each of these areas.

On December 5, 1985, Provost Brian J. Thompson asked me to serve as chairman of the review committee that he appointed to evaluate Paul Hunter's deanship and make recommendations for the future. During February and March of 1986, our committee conducted fifty interviews with individuals including all the chairs of the College of Arts and Science, selected faculty and students in the college, and three deans outside the College. On April 9, 1986, our committee reported to Provost Thompson (our entire report is included in my Correspondence Volume 1984–1991). The final paragraph gives a summary on our findings, "Viewed in its totality, Hunter's Deanship has had a very positive influence both on the College of Arts and Science and on the university community. He has worked to build strong departments and a distinguished faculty, to secure the financial health of the College, and to represent it well internally as well as at the national and international levels." Although we recommended that "a concerted effort be mounted to persuade Hunter to remain at the University of Rochester as Dean," he chose to return to academia and accepted a professorship in English at the University of Chicago. On June 27, 1986, Hunter wrote to me explaining his decision to leave and thanking me for my support and our report on his years as dean.

On July 21, 1986, I received a letter from Provost Brian J. Thompson stating, "On the basis of the recommendation of your faculty colleagues, I am pleased to ask you if you will serve as a member of the Search Committee for a new Dean of the College of Arts and Science." On August 29, Provost Thompson sent a letter to faculty announcing the names of the Search Committee members that he and President O'Brien had selected.

During this time period I was a member of the US Energy Research Advisory Board (ERAB), a board that advised the Secretary of Energy. I served under three different secretaries in the 1980s, D. P. Hodel, J. S. Herrington, and Admiral J. D. Watkins. ERAB advised the Secretary of Energy and the principal officers of the department on energy research and development programs and policies. "In this capacity the Board plays an important role in providing outside technical advice to the Department of Energy. The findings and recommendations of the Board have provided, and should continue to provide, valuable input to the decision making process that shapes our Nation's energy future."

ERAB met four times a year in the Forrestal Building in Washington DC, the home of the Department of Energy. Most of the Board's work was done by study groups. For example, I served on several time-consuming panels that produced the following reports: (1) Review of the NRC Report: Major Facilities for Materials Research and Related Disciplines (DOE/S-0037, 6/85); (2) Guidelines for DOE Long Term Civilian Development: Basic Energy Sciences, High Energy and Nuclear Physics (DOE/S-0045, 12/85); (3) Guidelines for DOE Long Term Civilian Research and Development: Overview and Summary (DOE/S-0046, 12/85); (4) Review of the National Research Council Report: Opportunities in Chemistry-Prepared by Chemistry Review Panel (DOE-0050, 5/86); (5) Review of NRC Report: Physics through 1990s—Prepared by the Physics Review Panel (DOE-0058, 2/87); and (6) Science and Engineering Education—Prepared by the Education Panel (DOE-0065, 7/88).

During the summer, ERAB held study sessions usually located at a national laboratory site. I remember especially the meetings at the Stanford Linear Accelerator and the Brookhaven National Laboratory. I chaired the ERAB Panel responsible for selecting the Fermi, Lawrence, and Renewable Energy awards for the DOE during the 1986–90 period. In 1984, I chaired a Review Panel for the Large Einsteinium Accelerator Program (LEAP) proposal submitted by four national laboratories to the DOE. During this time, I continued to be a member of the University of California President's Scientific and Educational Advisory Committee, meeting annually with President Saxon, the leader of the Research Universities of California. For a six-year period, 1985–1990, I was a member of the Physics Advisory Committee of the Oak Ridge National Laboratory.

On May 14, 1984, I joined my nuclear chemistry colleagues and members of the National Academy of Sciences, G. Friedlander, F. S. Rowland and G. T. Seaborg (Rowland and Seaborg are also Nobel Laureates) in alerting Dr. Alvin Trivelpiece, Director of Energy Research of DOE that the subfield of Nuclear and Radiochemistry has not had an appropriate home within the DOE (nor within NSF) for the past decade. We stated, "In our view this has had a negative impact on research in this subfield and has contributed to a growing shortage of chemists with nuclear science training. This shortage represents an erosion of an important national intellectual resource for vitally needed, often safety-related science and technology; we believe the country can ill afford such erosion, and this issue should be addressed by the Department of Energy." Having recently returned from a National American Chemical Society Meeting, I surmise that the field of Nuclear and Radiochemistry again faces similar funding problems in 2008!

During this five-year period (1983–88), I presented twenty-seven invited lectures on my research. Nineteen of these lectures were at national meetings and US universities. Among the US lectures were the DuPont Distinguished Lecture at Indiana University and the Joseph W. Kennedy Lectures at Washington University (one lecture to a general scientific audience and a second more specialized lecture). I mentioned earlier my invited lecture in 1984 at the Centenary of Niels Bohr in Copenhagen. An additional seven lectures were given abroad in Belgium, Greece, Japan, USSR, and France.

One of the surprises and true highlights of this period was the "Festschrift held on the occasion of the 65th Anniversary of my Birth" occurring on April 21, 1921. An International Symposium of Nuclear Fission and Heavy-Ion-Induced Reactions was organized by W. Udo Schröder and Hartwig Friesleben and held on April 20–22, 1986, with a large number of participants from the United States and abroad, many of whom were my former students, postdoctorals, and colleagues. In addition to the eight scientific sessions on April 21 and 22, there was a welcoming party on the evening of April 20, a reception at the residence of the Provost, and a banquet at the Rose Mansion on the evening of the 21st, and cocktails and dinner at the Wilson Commons on the 22nd. The Proceedings of the Conference were edited by W. U. Schröder and published in a 500-page book by Harwood Academic Publishers as part of their Nuclear Science Research Conference Series (Volume 11).

The subjects covered in the symposium were in research areas of particular interest to me, both past and present. It was a joyous occasion to be so honored by one's colleagues, and I greatly appreciate the work of Udo and Hartwig in organizing such an event and Niki Hansen Fowler's contribution in taking care of the many organizational details.

XI. The Years to Retirement (1988–1991)

During this period I returned full-time to research and teaching. A considerable portion of my time was diverted, however, to an investigation of a bizarre event that came to be known as "cold fusion." University of Utah scientists B. Stanley Pons and Martin Fleischmann claimed to have successfully created a sustained nuclear fusion reaction at room temperature in a small jar on a laboratory table-top. In 1989, I served as co-chair of a DOE/ERAB panel of twenty distinguished scientists to evaluate this extraordinary claim.

A. Heavy-Ion Reactions

Mass number, atomic number, energy, and angular distributions have been measured for projectile-like fragments from the reaction ^{238}U + ^{48}Ca at a laboratory energy of 425 MeV (246). The experimental atomic number (Z) distribution for all projectile-like fragments with energy losses greater than 20 MeV is shown in figure 45A. Correlations in the measured two-dimensional probability distributions P (A, Z) are discussed along with the general dissipative features of the damped reaction mechanisms. The first and second moments of the observed atomic and mass number distributions are compared with those of other asymmetric reaction systems. Although the ^{238}U + ^{48}Ca system is highly asymmetric, its potential energy surface in the vicinity of the injection point has a nearly zero gradient resulting in a negligible drift in the proton number. The proton drifts observed in reactions of ^{238}U with five different projectiles vary systematically with the potential gradient (see figure 45B). The charge and mass data are compared with quantitative predictions of a dynamical reaction model based on nucleon exchange. The theoretical model gives a reasonable account of the charge and mass distributions for the ^{238}U + ^{48}Ca reaction but fails systematically to predict experimental drifts when the driving forces are large.

Energy spectra of neutrons emitted in the damped reaction ^{139}La + ^{40}Ar at a laboratory energy or 400 MeV were measured at eight angles in coincidence with projectile-like reaction fragments, using a time-of-flight technique (247). Apart

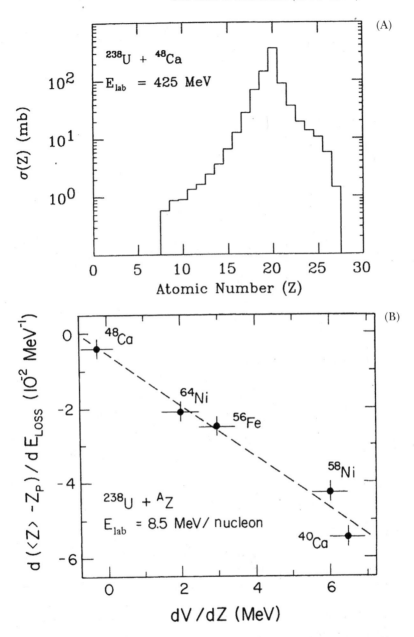

Figure 45. (A) Experimental atomic number (Z) distribution for all projectile-like fragments with energy losses greater than 20 MeV (246). (B) Correlation of the slope of the proton drift as a function of energy loss (for energy losses less than 50 MeV) with the average slope of the potential energy per Z unit determined for the vicinity of the injection point (246).

from a small high-energy component, the angular and energy distributions of neutrons were found to be well described by assuming two sources moving at velocities of the light and heavy reaction fragments emitting isotropically in their rest frames. The logarithmic shapes of the neutron energy spectra and the corresponding multiplicities of neutrons from the two sources suggest that the projectile-like fragment receives more than its equilibrium share of excitation energy for all values of the total kinetic energy loss. Detailed statistical calculations indicate that the amount of excitation energy generated in each fragment is well described by the one-body nucleon exchange model. The high-energy component in the neutron energy spectra has properties that are consistent with pre-equilibrium neutron emission. Averaged over all energy losses greater than 20 MeV, this "pre-equilibrium" neutron multiplicity is approximately equal to 0.1. However, the intensity of this component increases as the impact parameter decreases.

Energy and angular distributions of neutrons emitted in coincidence with projectile-like fragments from the damped reaction ^{139}La + ^{40}Ar at a bombarding energy of 600 MeV have been measured (253, 255). Components corresponding to sequential emission from two fully accelerated reaction fragments as well as pre-equilibrium neutron emission have been separated as a function of energy loss and fragment atomic number.

The redistribution of kinetic energy among the reaction partners of a strongly damped heavy-ion collision has been studied (248, 249). The partition of the total excitation energy between the fragments from the damped reactions ^{139}La + (400, 600 MeV) ^{40}Ar and ^{165}Ho + (403 MeV) ^{56}Fe has been deduced as a function of impact parameter, from measurements of energy spectra and multiplicities of neutrons emitted from correlated reaction fragments in flight. The experimental data are well reproduced by detailed simulation calculations of heat convection in the dinucleus and the eventual statistical decay of the reaction primaries.

The shapes of alpha-particle spectra from hot high-spin compound nuclei produced in energetic heavy-ion fusion reactions were analyzed within the framework of a statistical model (250, 252). For deformations of a magnitude given by the rotating liquid-drop model, alpha-particle spectra and effective barriers are insensitive to whether deformed or spherical nucleus transmission coefficients are used. It is important, however, to include the deformation dependence in the nuclear level density. Statistical model calculations reproduce charged-particle evaporation spectra from heavy-ion fusion reactions, where large deformations, high excitation energies, and angular momenta are involved.

A recent claim of significant correlations between the fragment mass and excitation energy at a fixed total kinetic energy loss in a damped reaction is re-examined (251). It is shown that correlations of similar character and strength to those reported result from incorrect assumptions made in the original analysis of kinematical coincidence data from the ^{165}Ho + ^{56}Fe reaction at 504 MeV. Based on Monte Carlo simulations of the experiment, it is concluded that within the uncertainties resulting from the experimental method, the data from the above reaction are consistent with a damped reaction mechanism, in which, for a fixed kinetic energy loss, the fragment excitation energy is uncorrelated with its mass.

The process of fission of a composite nucleus produced in heavy-ion collisions is characterized in terms of the kinetic energy, mass, and angular distributions of the fission fragments (254). Measurements of associated particles emitted prior to scission demonstrate that heavy-ion fission is an inherently slow (~10^{-20}–10^{-19} seconds) process that is dominated by overdamped collective motion in the exit channel. Experimental fission excitation functions are consistent with the fission barriers of the finite-range liquid-drop model.

It is shown that the finite resolution inherent in the kinematical coincidence method leads to systematic errors in the deduced (primary) physical quantities if the latter are calculated based on mass and linear momentum conservation equations alone (257). As an example, application of this method for measuring excitation energy of the fragments from damped reactions is reviewed. In such a case, finite resolution effects generate significant instrumental, or "background" correlations between the physical quantities reconstructed in a straightforward fashion, hence, if not accounted for, they may lead to the qualitative misinterpretation of the data. Experimental measures are discussed, which appear necessary in order to ensure proper accuracy of the finite resolution corrections. An alternate method of data analysis is presented that is much less susceptible to the finite resolution effects discussed.

Data from a kinematical coincidence experiment on the damped reaction ^{165}Ho + ^{74}Ge at 8.5 MeV/nucleon have been reanalyzed (258). Although the new analysis confirms the presence of some correlations between the excitation-energy division and its mass asymmetry, the magnitude of these correlations is found to be significantly smaller than that previously reported. Proton-neutron symmetry of the mechanism of heat generation through nucleon exchange is revealed and a possible acceptor-donor asymmetry of the mechanism is discussed.

Neutron spectra from the inverse-kinematics reaction ^{58}Ni + ^{208}Pb at E_{lab}/ A= 6.65 MeV have been measured in coincidence with nickel-like fragments (259). Neutron emission patterns for net pickup and stripping channels have been analyzed in terms of sequential evaporation from fully accelerated projectile-like and target-like fragments. As at higher energies, these patterns suggest an absence of appreciable correlations between net mass transfer and excitation energy division for strongly damped collisions, at the present near-barrier energy. The overall multiplicities, as well as energy spectra and angular distribution of neutrons, are well reproduced by simulation calculations, assuming an energy division always in favor of the heavy fragment.

B. Other Activities

The high linear energy transfer of alpha particles makes them ideal for use in therapy. The path length in tissue is 60 or 100 micrometers, which minimizes the dose to non-target tissue. The nuclide ^{211}Pb and its daughters are produced in a generator system that utilizes the distillation of the intermediate daughter ^{219}Rn from ^{223}Ra (256). The radium is precipitated as the stearate to isolate the parent while allowing the gaseous daughter to emanate. While the yield of the system is low, the radionuclidic purity is extremely high.

In June, 1988, I was appointed to serve another three-year term as Chairman of the NAS Committee on Nuclear and Radiochemistry. This committee met monthly at the NAS building in Washington. During the summer of 1988, I chaired a session at the Gordon Conference on Nuclear Chemistry and also chaired a session at the Third International Conference on Nucleus-Nucleus Collisions held at Saint-Malo, France. In early October, 1988, I was in Darmstadt serving on the review committee of GSI, the laboratory housing Germany's heavy-ion accelerator facilities. I served on the Advisory Committee for the International Conference, "Fifty Years Research in Nuclear Fission" held in Berlin on April 3–7, 1989. In addition, I served on the planning committee for a second such conference held on April 26–28 at the National Academy of Science and the National Institute of Standards and Technology in Washington, DC Near the back of my Reference Volume labeled "Photos" are the signatures of those attending the historic meeting in Berlin.

On March 23, 1989, a University of Utah press release stated that Martin Fleischmann and B. Stanley Pons had successfully created a sustained nuclear fusion reaction at room temperature in a chemistry laboratory. Soon thereafter, Admiral Watkins, the Secretary of Energy, requested the Energy Research Advisory Board, a standing committee that advises the secretary on all matters of energy policy, to assemble a panel to assess "cold fusion." The chairman of ERAB then contacted me with a persuasive request to chair the panel. I had been a member of ERAB for several years and agreed under the condition that a co-chair of the panel also be appointed to share the workload. This in hindsight turned out to be a mistake. Norman Ramsey, Higgins Professor of Physics at Harvard, consented to serve as co-chair. However, he served only during the initial months, through the writing of the Interim Report. Hence, Ramsey did not participate in either the panel's further major studies, investigations, and deliberations or the preparation and writing of the final report. He returned on the last one-half day of our panel's open meeting in Washington to approve the final report and produced a short crisis. His attempt to weaken our panel's conclusions, however, was unanimously rejected by the other panel members.

The final report of our panel DOE/S-0073 entitled "Cold Fusion Research" was issued in November, 1989. In my letter to the chairman of ERAB, I stated, "I am pleased to forward to you the Final Report of the Cold Fusion Panel. This report reviews the current status of cold fusion and includes major chapters on Calorimetry and Excess Heat, Fusion Products and Materials Characterization. In addition, the report makes a number of conclusions and recommendations, as requested by the Secretary of Energy. The Panel or subgroups thereof have participated in the Workshop on Cold Fusion in Santa Fe, have visited several laboratories, have examined many published articles and preprints, studied numerous communications and privately distributed reports, and have participated in many discussions. In addition, the Panel held five public meetings where its findings were discussed and drafts of both the Interim and Final Reports were formulated." The Panel on "Cold Fusion" kept me extremely busy in 1989, and I followed the activity in the field worldwide for several more years. During the three-year period up until my retirement on June 30, 1991, I gave some twenty colloquia mostly on the subject of "Cold Fusion." I was getting requests for such talks from groups all over the United States. As an example, I fulfilled a request from the Cosmos Club in Washington, DC, to speak on "Cold Fusion" on March 15, 1990. In the audience of dignitaries was Allan Wallis, president of the University of Rochester at the time I was hired. After retirement from Rochester, he became an assistant to Secretary of State Schultz. It was a pleasant surprise to see him at the lecture.

On March 29–31, 1990, I attended the First Annual Conference on Cold Fusion held in Salt Lake City. Most of the more than 200 participants were believers and all the talks were positive. It was largely a US production with a token number of foreign papers and participants. Organizers tried to control the press. It was a meeting like no other scientific meeting that I've ever attended. At this conference I had a most unusual and bizarre exchange with Fleischmann. During a coffee break I approached him stating that one year had passed and since the expected fusion products had not been observed, he must have some hunch about the source of the heat that he was reporting. Without missing a beat, he responded that he thought the palladium was fissioning. Being an expert in fission physics and knowing the threshold for the fission of palladium was some tens of millions of electron volts, I lost the little confidence I had that Fleischmann knew what he was doing.

I was a member of the Oak Ridge National Laboratory Physics Advisory Committee during 1985–90, a member of the Chemistry Review Committee of the Lawrence Livermore National Laboratory for the period 1988–93, a member of an Advisory Panel for the Sandia National Laboratory from 1988–92, and a member of the Visiting Committee for the Brookhaven National Laboratory Chemistry Department form 1990–92.

On February 22 and 23, 1991, I attended the fiftieth anniversary of the discovery and first identification of plutonium at the Lawrence Berkeley Laboratory. A few of the participants were "old timers" who had worked on the Manhattan Project in the early forties. A group photo of the attendees can be found in my Reference Volume labeled "Photos." On March 6, 1991, I traveled to Bethpage, New York (on Long Island), the home of the Grumman Corporation, to receive the Leroy Randle Grumman Medal, along with a cash award, and to give a lecture entitled The Status of Cold Fusion Claims. Imprinted on the back of the ¼-inch thick by 3-inch diameter medal is the citation "Awarded by the Grumman Corporation for Outstanding Achievement to John R. Huizenga, March 6, 1991." This award and lecture series was established "to commemorate Leroy Grumman, the innovative aeronautical engineer who, as one of its founders, gave the Grumman Corporation its name and helped to shape its proud heritage."

Reaching my 70th birthday on April 21, 1991, I was required by the then existing law, which was soon after rescinded, to officially retire on June 30, 1991. On May 23, 1991, I received a letter from the Dean of the College of Arts and Science stating, "At the final faculty meeting of each academic year, it is customary for a minute to be read on the retirement of a member of the college faculty, I am pleased to enclose a copy of the minute on your retirement." A copy of this minute is enclosed in my Correspondence Volume 1984–91. My change of status to Tracy H. Harris Professor Emeritus also arrived from the President following a meeting of the Board of Trustees. The Chemistry Department retirement dinner was held on June 8, 1991, marking the end of my official career.

XII. The Years after Retirement (1991–)

For some years I continued to serve on review and advisory panels at several different national laboratories. In 1991, I was appointed to a multi-year term on the Advisory Council of the Glenn T. Seaborg Institute for Transactinium Science at the Lawrence Livermore National Laboratory. A group photo of the members of this advisory council is included in my Reference Volume labeled "Photos."

In the fall of 1991, Udo Schröder and I were informed that we were successful in our application to the Department of Energy's University Research Instrumentation (URI) program and recommended for funding in the amount of $300,000. These funds were used in the construction of a 4-π Neutron Multiplicity Meter (NMM) for calorimetric studies of Intermediate Energy Heavy Ion Reactions. Of the 184 applications for instrument grants, we were one of the fourteen grants awarded, receiving the second largest award. The reviewers' evaluation of our proposal and a list of the other instrumentation proposals are included in my Correspondence Volume 1984–1991.

During the years after retirement I continued to examine the numerous "cold fusion" claims coming from both abroad and within the United States. The intense interest worldwide in the subject motivated me to record my experience in a book (260) published in 1992. This book entitled, *Cold Fusion: The Scientific Fiasco of the Century*, was published by the University of Rochester Press and included events occurring up to June 30, 1991, when the National Fusion Institute closed. This was, however, by no means the end of the cold fusion fiasco. Claims of successful cold fusion experiments continued to appear in media stories, preprints, conference proceedings, and peer-reviewed journals. In fact, some of the more recent events associated with cold fusion could even be classified as downright sensational. Take, for example, the several new reports claiming enormous amounts of excess heat with ordinary *light* water! Others postulated element transmutation in the chemistry laboratory, including even the preposterous claim of making gold! Hence, a revised and updated paperback edition (266) was published in 1993 by Oxford University Press. The paperback edition was reprinted in 1994 including events such as the Fourth International

Conference on Cold Fusion held on December 6–9, 1993, in Maui. In both the hardback and the paperback editions a chronology of the cold fusion saga is included in Appendix III.

In addition to the First International Conference on Cold Fusion mentioned earlier, I also attended the Third Conference held in Nagoya, Japan (10/21–25/1992) and the Fourth Conference held in Maui (December 6–9, 1993). The cold fusion proponents excelled in organizing conferences in exotic locations and luxurious hotels. The most striking thing that happened in Nagoya was the stock market response to a positive cold fusion claim made in a press conference by E. Yamaguchi, a scientist at the Nippon Telegraph and Telephone (NTT) Corporation. On the basis of his unverified experiments, NTT's stock rose 11 percent amounting to a paper profit of approximately 8 billion dollars.

While in Japan, I visited several universities and learned firsthand that mainline scientists in Japan had the same negative opinion about cold fusion as did the mainline scientists in the United States. The most notable thing about the Fourth International Conference was that it was held in the luxurious Hyatt Regency Hotel. The good news at this conference was an announcement by Kazuski Matsui that the Ministry of International Trade and Industry had organized a New Hydrogen Energy research project in cooperation with key Japanese companies and universities with a proposed budget equal to 30 million dollars over the next four years! The bad news was that after fifty-six months of activity, there was no persuasive evidence presented in support of cold fusion!

Shortly before my first book on cold fusion was to be available, I received a letter from Attorney Patrick A. Shea, representing Pons and Fleischmann, stating, "My clients have expressed concern that your book could contain statements that would be considered defamatory. This letter shall serve as notice upon you that any such defamatory statements would be vigorously pursued on the part of Professor Pons and Professor Fleischmann and therefore a review prior to publication would be appreciated. With, or without a review, should we determine your publication contains defamatory or other actionable material we will immediately seek appropriate legal recourse." This was a blatant attempt to delay or stop publication of my book by threatening me with expensive litigation! Shea's letter indeed served to scare me. This fear, however, was abated after Robert Easton, University of Rochester Press, and I consulted with John J. Diehl, University of Rochester Counsel. After reviewing my book, Diehl wrote us on April 20, 1992, and in part stated, "Because you have received correspondence from an attorney claiming to represent Pons and Fleischmann in matters of libel, you have each expressed concern that the University be prepared to stand behind the book and defend it against any false libel claims. Since the book is not defamatory, we don't think it will come to that, but rest assured that the University will provide such a defense, should that become necessary."

In addition to the books on cold fusion, I also wrote articles on the subject for the Forum for Applied Research and Public Policy (261) and for Leaders Magazine (263). During this time period, I gave some thirty invited lectures on cold fusion including a plenary lecture at national meetings of the American Physical Society, the American Chemical Society, and the American Association for the Advancement of Science. Included also were lectures in Tokyo and Copenhagen.

In 1992, I was inducted into the American Academy of Arts and Sciences. The citation stated, "John Huizenga has significantly influenced the progress of nuclear science through his discovery of the hidden fundamental principles and systematics of very complex physical processes, such as fission and heavy-ion-induced reactions."

As an aside, election into the American Academy sixteen years after election into the more selective and prestigious National Academy was a bit anticlimactic, to say the least. The usual election sequence for scientists is election into the American Academy years prior to election, if at all, into the National Academy.

Neutrons from the fusion reactions of 120 to 150 MeV ^{28}Si with ^{118}Sn and ^{124}Sn target nuclei have been measured in coincidence with evaporation residues in two series of complementary experiments using either a 4π neutron multiplicity meter or a neutron time-of-flight spectrometer (262). Both energy spectra and multiplicity distributions reveal significant quantitative differences in the decay patterns of the compound nuclei ^{146}Gd and ^{152}Gd formed in the two reactions studied. These differences cannot be understood in terms of decay cascades proceeding through states of enhanced collective energy, such as the super deformed states suggested in earlier studies. Instead, they can be explained consistently within the framework of a statistical decay model, if different effective level density parameters are allowed for the evaporation chains of the two composite systems.

In 1994, three years after my official retirement, I was still actively involved full time in a variety of scientific activities including research, committee and advisory panel assignments, and in debunking "cold fusion" claims. However, because of Dolly's failing health, we made a decision to leave the snow and cold climate of the Rochester winters. We moved south to Pinehurst, North Carolina, a small resort community in the central part of the state, noted for its many golf courses. Pinehurst is somewhat more than one hour south of the Research Triangle that is surrounded by three major universities: Duke, University of North Carolina, and North Carolina State University.

We built a home on the 17th fairway, near the green, of the Magnolia Course. The house is French Traditional style and won the 1995 Grand Prize of the Moore County Home Builders Association. The 17th hole is a long par-4 with hundreds of tall pines, magnolias, dogwoods, and flowering bushes lining the fairway. Construction took nine months and we moved into our new home on July 3, 1995. I can sit at my desk in the library and see the action on the green as well as most of the fairway.

On January 20, 1995, I had a St. Jude aortic valve implant and a triple bypass performed at Cedars-Sinai Medical Center in Los Angeles. I took advantage of my doctor-son Rob's expertise and his knowledge of the surgeons at this medical center. His medical practice is located in Beverly Hills and he sends many of his patients to Cedars-Sinai.

After recovery, I began to master the game of golf at the young age of 74. I enjoyed being on the golf course with its natural beauty; however, my game improved only slowly. Even so, I enjoyed the game sufficiently to play several rounds a week for several years. Teamed with Jack Lund, a low-handicap golfer, we won our club's Blazer tournament in 1999. This is a two-man best-ball tournament with sixty-four entering teams. Each of the players on the winning team receives a blazer. Since golf is a handicapped game, players with a wide range of abilities can compete with each other.

Another highlight of my golfing career occurred on August 3, 2006, when I had a hole-in-one on the 17th hole of our Holly course. During this time period I also had the opportunity to serve as a marshall at two US Open tournaments held on Pinehurst's famous Number 2 course, in 1999 and 2005. Seeing at close range the professional golfers play is awesome. Tiger Woods and Mark O'Mara rented a house directly across the fairway from our home and, in the evening, I could watch them practice their putting from my library window. The 1999 US Open tournament was especially exciting with the Payne Stewart win after making a long putt on the final hole! (Sadly, a couple of months later, he was carried to his death on a runaway Learjet, which had flown out of control halfway across the United States before nosediving into a field in South Dakota.) Pinehurst has erected a bronze statue of Stewart near the clubhouse.

During the first week of April, 1996, we held an open house to celebrate our fiftieth wedding anniversary. We sent invitations to friends to come when you can and we'll supply the food, fun, and fellowship. All of our children and grandchildren came for the week. A large number of friends also came from near and far during the week.

On June 3, 1996, I was invited to the Argonne National Laboratory to deliver a joint Chemistry Division–Physics Division seminar in celebration of Argonne's fiftieth anniversary. I was assigned the topic "The Discovery of Elements 99 and 100: Prelude to a Rich History of Accomplishments in Nuclear Chemistry at Argonne." I was indeed honored to participate in this historic celebration and chose to discuss four areas of research, all of which I had participated in during my eighteen years at Argonne.

The first and principal topic I discussed was our discovery and characterization of elements 99 and 100, as well as many other new transplutonium nuclides (see Chapter II). These nuclides were prepared by long irradiations of ^{239}Pu in the high neutron flux Materials Testing Reactor in Idaho and from the collection and analyses of debris from the "Mike" thermonuclear explosion in 1952. Emphasis was placed in my talk on some of the key discovery experiments and the important role played in making our elemental assignments on our previously acquired understanding of the systematics of spontaneous fission half-lives.

The second topic covered was cosmochemistry, including the measurements of elemental and isotopic abundances by neutron irradiation of chondritic meteorites (see Subsection IV.A). I discussed the astrophysical "r" and "s" neutron-capture processes coined to represent neutron capture on a rapid- and slow-time scale, respectively. These two very different neutron-capture processes have their origins in our terrestrial experiments, where "Mike" demonstrated 17th order neutron capture on ^{238}U in a fraction of a second, and the MTR experiments, over a long period of time, produced massive neutron capture along the beta-stability line. The "r" and "s" neutron-capture processes have each contributed in their own unique way to the present-day observed elemental abundances.

The third topic covered was nuclear fission research (see Chapter II, Subsection IV.B, and Chapter VI). Experiments on fission included measurements of cross sections, angular and kinetic energy distributions, as well as detailed studies of the competition between neutron emission and fission. Some emphasis was placed on our pioneering studies of the low-energy levels in the transition-state nucleus by the $(\alpha,\alpha'f)$ reaction.

The last subject discussed was nuclear reactions and spectroscopy, including the discovery of a very low-energy isomer in the well-known fissile nuclide ^{235}U. This explained a long-time puzzle in the alpha decay of ^{239}Pu, another well-known fissile nuclide.

During the last years of Dolly's life, she was on oxygen around the clock. We kept forty-liter containers of liquid oxygen in the house at all times. She was able to leave the house for short periods of time to play bridge or attend a movie by taking along one or more small oxygen tanks. She especially enjoyed art movies. It was in the 1980s that Dolly began to have serious respiratory problems. She was diagnosed at Strong Hospital to be suffering from asthma and allergies and an intensive, but failing, effort was made to identify the causative substances.

With Dolly's condition worsening and getting little satisfaction from her doctor, we again took advantage of Rob's expertise. Following a trip to La Jolla in February 1993 to see Joel, we next visited Rob, Wanda, and family in Los Angeles during March 1993. Rob immediately placed Dolly in the Cedars-Sinai Medical Center and in consultation with staff ordered a number of diagnostic procedures, including chest X-rays and a lung biopsy. The experts at Cedars-Sinai conclusively diagnosed her with pulmonary interstitial fibrosis, acute and chronic. Unfortunately, the correct diagnosis came rather late and one can only speculate about the outcome if her ailment had been diagnosed earlier. So much for asthma and allergies! However, at this time we were placing our home in Rochester on the market and planning our departure for Pinehurst.

Litigation was going to be time consuming, and we were not prepared to put the necessary time and energy into a legal fight.

Nevertheless, we felt it important to express our unhappiness with the way one of Dolly's doctors in Rochester managed her medical care. On August 19, 1994 (after we were in Pinehurst), Dolly wrote to him and stated (in part), "under your care, I was misdiagnosed over a five-year period as having asthma and allergies due to the house dust mite. My diagnosis at Cedars-Sinai was interstitial fibrosis, acute and chronic . . . not bronchial asthma for which you were treating me with inhalers and allergy shots since 1988! As a result, the quality of my life has deteriorated markedly and I have lost one-half of my lung capacity, now requiring oxygen. To our great surprise on seeing the Rochester medical records at Cedars-Sinai, we learned that I had a pulmonary function test (PFT) in early 1988 and that the report stated 'This could be the beginning of interstitial lung disease.' You chose to ignore this warning and never followed up with another PFT, chest x-ray or sedentary rate test over the next five years! The whole episode appears to be a case of tunnel vision—seeing nothing beyond asthma and allergies—even though you had no solid evidence to support this diagnosis."

After this long battle with pulmonary fibrosis, Dolly died on May 1, 1999.

My entire collection of cold fusion materials was donated in June, 2003, to the Department of Rare Books, University of Rochester Libraries. This collection includes several copies of our panel's interim and final reports, several books, correspondence, newspaper and journal articles, conference proceedings and materials, e-mails, photographs, and press releases collected during 1989–2003. The library staff organized the material into categories labeled (1) Reports, (2) Conferences, (3) Articles, (4) Articles (con't) (5) Other Materials, and (6) Other Materials—miscellaneous. Of all the major newspapers reporting on cold fusion, the *Wall Street Journal* displayed the least regard for objective reporting. It completely ignored the opinions of mainline scientists and published some fifty favorable stories about "Cold Fusion"! This series of articles, appearing over a period of several years, and included in the above-mentioned library collection, holds a world record for its blatant disregard for the views of mainline scientists.

On April 21, 2006, the University of Rochester sponsored a Nuclear Science Symposium in honor of my 85th birthday. During the day a series of lectures on current topics in nuclear science were presented by former students, research associates, and colleagues. In the evening, a reception and gala banquet was held at the Memorial Art Gallery. It was a memorable event for me with so many former research collaborators and colleagues in attendance. Many remained for an informal brunch on Saturday.

It was at the above birthday celebration that Debra Haring suggested to me that I write about events that had occurred during my career. In late 2007, I began to write such a summary of my earlier research activities, including numerous related activities associated with my five-decades-long career in nuclear science. Herewith are the results, which likely speak for themselves.

This memoir, along with the archive of my technical publications and correspondence, do not give the meaning to such a career, but they do provide the paper trail of a life well-lived—as scientist, advisor, and beloved son, husband, father, and grandfather.

Appendix

1.   John R. Huizenga, Philip F. Greiger, and Frederick T. Wall
     Electrolytic properties of aqueous solutions of polyacrylic acid and sodium
     hydroxide. I. Transference experiments using radioactive sodium
     J. Am. Chem. Soc. 72, 2636 (1950)
2.   John R. Huizenga, Philip F. Greiger, and Frederick T. Wall
     Electrolytic properties of aqueous solutions of polyacrylic acid and sodium
     hydroxide. II. Diffusion experiments using radioactive sodium
     J. Am. Chem. Soc. 72, 4228 (1950)
3.   J. R. Huizenga, L. B. Magnusson, O. C. Simpson, and G. H. Winslow
     Neutron binding energies
     Phys. Rev. 79, 908 (1950) (letter)
4.   J. R. Huizenga, L. B. Magnusson, P. R. Fields, M. H. Studier, and R. B.
     Duffield
     Threshold for photoneutron reaction in ^{238}U
     Phys. Rev. 82, 561 (1951) (letter)
5.   John R. Huizenga and Lawrence B. Magnusson
     Oxidation-reduction reactions of neptunium (IV) and–(V)
     J. Am. Chem. Soc. 73, 3202 (1951)
6.   L. B. Magnusson, J. R. Huizenga, P. R. Fields, M. H. Studier, and R. B.
     Duffield
     Threshold for photoneutron reaction in ^{232}Th
     Phys. Rev. 84, 166 (1951) (letter)
7.   J. R. Huizenga, L. B. Magnusson, M. S. Freedman, and F. Wagner Jr.
     The binding energy of four neutrons in ^{239}U and the disintegration ener-
     gies of ^{239}U and ^{239}Np
     Phys. Rev. 84 1264 (1951) (letter)
8.   J. C. Brantley and J. R. Huizenga
     Solubility of cerium (IV) pyrophosphate
     J. Am. Chem. Soc. 74, 6101 (1952)
9.   Frederick T. Wall, Philip F. Greiger, John R. Huizenga, and Robert H.
     Doremus
     Electrolytic properties of aqueous solutions of polyacrylic acid and sodium
     hydroxide. III. The rate of sodium ion exchange between polyacrylate
     and free sodium ions
     J. Chem. Phys. 20, 1206 (1952)

10.   John R. Huizenga and Robert B. Duffield
       Fission-to-capture cross-section ratio
       Phys. Rev. 88, 959 (1952) (letter)

11.   F. Wagner, Jr., M. S. Freedman, D. W. Engelkemeir, and J. R. Huizenga
       Radiations of 6.7-day ^{237}U
       Phys. Rev. 89, 502 (1953)

12.   L. B. Magnusson and J. R. Huizenga
       Stabilities of +4 and +5 oxidation states of the actinide elements–the Np
       (IV)–Np (V) couple in perchloric acid solution
       J. Am. Chem. Soc. 75, 2242 (1953)

13.   Robert B. Duffield and John R. Huizenga
       Photofission and photoneutron emission in uranium
       Phys. Rev. 89, 1042 (1953)

14.   D. W. Engelkemeir, P. R. Fields, and J. R. Huizenga
       Radiations of ^{243}Pu
       Phys. Rev. 90, 6 (1953)

15.   J. R. Huizenga, W. M. Manning, and G. T. Seaborg
       Slow neutron and spontaneous fission properties of heavy nuclei
       Division IV, Volume 14A, Chapter 20, pp. 839–853, *The Actinide Elements,*
       Glenn T. Seaborg and Joseph J. Katz, editors, National Nuclear Energy
       Series (McGraw-Hill Book Co., New York, 1954)

16.   M. H. Studier, P. R. Fields, H. Diamond, J. F. Mech, A. M. Friedman, P. A.
       Sellers, G. Pyle, C. M. Stevens, L. B. Magnusson, and J. R. Huizenga
       Elements 99 and 100 from pile irradiated plutonium
       Phys. Rev. 93, 1428 (1954) (letter)

17.   P. R. Fields, M. H. Studier, and J. R. Huizenga
       Will actinide series be completed?
       Chem. and Eng. News 93, 1213 (1954)

18.   M. H. Studier, P. R. Fields, P. A. Sellers, A. M. Friedman, C. M. Stevens, J.
       F. Mech, H. Diamond, J. Sedlet, and J. R. Huizenga
       Plutonium-244 from pile irradiated plutonium
       Phys. Rev. 93, 1433 (1954) (letter)

19.   J. R. Huizenga
       Spontaneous fission systematics
       Phys. Rev. 94, 158 (1954)

20.   P. R. Fields, M. H. Studier, J. F. Mech, H. Diamond, A. M. Friedman, L. B.
       Magnusson, and J. R. Huizenga
       Additional properties of isotopes of elements 99 and 100
       Phys. Rev. 94, 209 (1954) (letter)

21.   C. M. Stevens, M. H. Studier, P. R. Fields, J. F. Mech, P. A. Sellers, A. M.
       Friedman, H. Diamond, and J. R. Huizenga
       Curium isotopes 246 and 247 from pile irradiated plutonium
       Phys. Rev. 94, 974 (1954) (letter)

22. H. Diamond, L. B. Magnusson, J. F. Mech, C. M. Stevens, A. M. Friedman, M. H. Studier, P. R. Fields, and J. R. Huizenga
Identification of californium isotopes 249, 250, 251, and 252 from pile irradiated plutonium
Phys. Rev. 94, 1083 (1954) (letter)

23. P. R. Fields, M. H. Studier, L. B. Magnusson, and J. R. Huizenga
Spontaneous fission properties of elements 97, 98, 99, and 100
Nature 174, 265 (1954)

24. J. R. Huizenga, J. E. Gindler, and R. B. Duffield
Relative photofission yields of several fissionable materials
Phys. Rev. 95, 1009 (1954)

25. A. M. Friedman, A. L. Harkness, P. R. Fields, M. H. Studier, and J. R. Huizenga
Alpha half-lives of ^{244}Cm, ^{245}Cm, AND ^{246}Cm
Phys. Rev. 95, 1501 (1954)

26. M. H. Studier, and J. R. Huizenga
Correlation of spontaneous fission half-lives
Phys. Rev. 96, 545 (1954) (letter)

27. J. R. Huizenga and C. M. Stevens
New long-lived isotopes of lead
Phys. Rev. 96, 548 (1954) (letter)

28. L. B. Magnusson, M. H. Studier, P. R. Fields, C. M. Stevens, J. F. Mech, A. M. Friedman, H. Diamond, and J. R. Huizenga
Berkelium and californium isotopes produced in neutron irradiation of plutonium
Phys. Rev. 96, 1576 (1954)

29. P. R. Fields, J. E. Gindler, A. L. Harkness, M. H. Studier, J. R. Huizenga, and A. M. Friedman
Electron capture decay of ^{244}Am and the spontaneous fission half-life of ^{244}Pu
Phys. Rev. 100, 172 (1955)

30. A. Ghiorso, S. G. Thompson, G. H. Higgins, G. T. Seaborg, M. H. Studier, P. R. Fields, S. M. Fried, H. Diamond, J. F. Mech, G.L. Pyle, J. R. Huizenga, A. Hirsch, W. M. Manning, C. I. Browne, H. L. Smith, and R. W. Spence
New elements einsteinium and fermium, atomic numbers 99 and 100
Phys. Rev. 99, 1048 (1955) (letter)

31. J. R. Huizenga
The nuclear fission process
Proc. Int. Conf. on Peaceful Uses of Atomic Energy, Paper No. 836, *2*, 208–213 (1956)

32. W.C. Bentley, H. Diamond, P. R. Fields, A. M. Friedman, J. E. Gindler, D. J. Hess, J. R. Huizenga, M. G. Inghram, A. M. Jaffey, L. Magnusson, W. M. Manning, J. F. Mech, G.L. Pyle, R. Sjoblom, C. M. Stevens, and M. H. Studier

The formation of higher isotopes and higher elements by reactor irradiation of ^{239}Pu. Some nuclear properties of the heavier isotopes
Proc. Int. Conf. on Peaceful Uses of Atomic Energy, Paper No. 809, 7, 261–273
(1956)

33.   J. R. Huizenga
      Isotopic masses III. A > 201
      Physica XXI, 410 (1955)

34.   C. Andre, J. Huizenga, J. Mech, W. Ramler, E. Rauh, and S. Rocklin
      Proton cross sections of ^{209}Bi
      Phys. Rev. 101, 645 (1956)

35.   J. R. Huizenga and J. Wing
      Long lived lead-205
      Phys. Rev. 102, 926 (1956) (letter)

36.   J. R. Huizenga, V. E. Krohn, and S. Raboy
      Angular correlation of gamma rays in ^{204}Pb
      Phys. Rev. 102, 1063 (1956)

37.   S. M. Fried, G. Pyle, C. M. Stevens, and J. R. Huizenga
      Spontaneous fission half-life of curium-246
      J. Inorg. and Nucl. Chem. 2, 415 (July, 1956) (letter)

38.   J. F. Mech, H. Diamond, M. H. Studier, P. R. Fields, A. Hirsch, C. M. Ste-
      vens, R.F. Barnes, D.J. Henderson, and J. R. Huizenga
      Alpha and spontaneous fission half-lives of plutonium-242
      Phys. Rev. 103, 340 (1956)

39.   R. H. Herber, T. T. Sugihara, C.D. Coryell, W. E. Bennett, and J. R. Huiz-
      enga
      Probable Absence of K Capture in the Decay of Lead-205
      Phys. Rev. 103, 955 (1956)

40.   J. E. Gindler, J. R. Huizenga, and R.A. Schmitt
      Photofission and photoneutron emission in thorium and uranium
      Phys. Rev. 104, 425 (1956)

41.   J. R. Huizenga and J. Wing
      Radioactivity of ^{44}Ti
      Phys. Rev. 106, 90 (1957)

42.   J. R. Huizenga, C. L. Rao, and D. W. Engelkemeir
      27-minute isomer of ^{235}U
      Phys. Rev. 107, 319 (1957)

43.   J. R. Huizenga and H. Diamond
      Spontaneous fission half-lives of ^{254}Cf AND ^{250}Cm
      Phys. Rev. 107, 1087 (1957)

44.   G. L. Bate, J. R. Huizenga, and H. A. Potratz
      Thorium content of stone meteorites
      Science 126, 612 (1957)

45.   J. R. Huizenga
      Correlation of the competition between neutron emission and fission
      Phys. Rev. 109, 484 (1958)

46.   J. E. Gindler, J. R. Huizenga, and D. W. Engelkemeir
      Neptunium isotopes: 234, 235, 236
      Phys. Rev. 109, 1263 (1958)

47.   F.A. White, F.M. Rourke, J.C. Sheffield, R.P. Schuman, and J. R. Huizenga
      Absolute measurement of alpha particles from ^{210}Po
      Phys. Rev. 109, 437 (1958)

48.   J. R. Huizenga
      Competition between fission and neutron emission in heavy nuclei
      Proc. Conf. on Reactions between Complex Nuclei, ORNL-2606 (1958)

49.   George L. Bate, Herbert A. Potratz, and J. R. Huizenga
      Thorium in iron meteorites: a preliminary investigation
      Geochim. Et Cosmochim. Acta 14, 118 (1958)

50.   J. Wing, C. M. Stevens, and J. R. Huizenga
      Radioactivity of lead-205
      Phys. Rev. 111, 590 (1958)

51.   R. Vandenbosch and J. R. Huizenga
      Nuclear fission process: a study of the competition between fission and
      neutron emission as a function of excitation energy and nuclear type
      Proc. Int. Conf. Peaceful Uses of Atomic Energy, 2nd 15, 284 (1958), United
      Nations, Geneva

52.   DC Hess, G. W. Reed, and J. R. Huizenga
      Whence earth?
      Ind. And Eng. Chem. 50, 26A (1958)

53.   G. J. Nijgh, A. H. Wapstra, L.Th.M. Ornstein, N. Solomons-Grobben, and
      J. R. Huizenga
      Conversion coefficients of gamma transitions in ^{203}Tl
      Nucl. Phys. 9, 528 (1958–59)

54.   G. L. Bate, J. R. Huizenga, and H. A. Potratz
      Thorium in stone meteorites by neutron activation analysis
      Geochim. Et Cosmochim. Acta 16, 88 (1959)

55.   J. E. Gindler, J. Gray, Jr., and J. R. Huizenga
      Neutron fission cross sections of ^{236}Pu and ^{237}Pu
      Phys. Rev. 115, 1271 (1959)

56.   W. J. Ramler, J. Wing, D. J. Henderson, and J. R. Huizenga
      Excitation functions of bismuth and lead
      Phys. Rev. 114, 154 (1959)

57.   J. Wing, W. J. Rambler, A. L. Harkness, and J. R. Huizenga
      Excitation functions of ^{235}U and ^{238}U bombarded with helium and deu-
      terium ions
      Phys. Rev. 114, 163 (1959)

58. W. D. Ehmann and J. R. Huizenga
     Bismuth, thallium, and mercury in stone meteorites by activation analysis
     Geochim. et. Cosmochim. Acta 17, 125 (1959)

59. G. L. Bate, A. H. Potratz, and J. R. Huizenga
     Scandium, chromium, and europium in stone meteorites by simultaneous neutron activity analysis
     Geochim. et. Cosmochim. Acta 18, 101 (1960)

60. J. R. Huizenga and R. Vandenbosch
     Interpretation of isomeric cross-section ratios for (n, Γ) and (Γ, n) reactions
     Phys. Rev. 120, 1305 (1960)

61. R. Vandenbosch and J. R. Huizenga
     Isomeric cross-section ratios for reactions producing the isomeric pair ^{197}Hg and ^{197m}Hg
     Phys. Rev. 120, 1313 (1960)

62. W. D. Ehmann and J. R. Huizenga
     A search for long-lived ^{50}Ca and ^{56}Cr
     Transactions of the Kentucky Academy of Science 21 (No. 1–2), 1 (1960)

63. R. Vandenbosch, J. R. Huizenga, W. F. Miller, and E. M. Keberle
     Monte Carlo calculation of neutron evaporation: excitation energy dependence of nuclear level density
     Nucl. Phys. 25, 511 (1961)

64. J. Wing, B. A. Swartz, and J. R. Huizenga
     New hafnium isotope, ^{182}Hf
     Phys. Rev. 123, 1354 (1961)

65. R. Vandenbosch, H. Warhanek, and J. R. Huizenga
     Fission fragment anisotropy and pairing effects on nuclear structure
     Phys. Rev. 124, 846 (1961)

66. J. R. Huizenga, R. Vandenbosch, and H. Warhanek
     Reaction cross sections of ^{233}U and ^{238}U with 18–43 MeV helium ions
     Phys. Rev. 124, 1964 (1961)

67. J. R. Huizenga and G. Igo
     Theoretical reaction cross sections for alpha particles with an optical model
     Nucl. Phys. 29, 462 (1962)

68. J. R. Huizenga, R. Chaudhry, and R. Vandenbosch
     Helium-ion-induced fission of Bi, Pb, Tl, and Au
     Phys. Rev. 126, 210 (1962)

69. R. Chaudhry, R. Vandenbosch, and J. R. Huizenga
     Fission fragment angular distributions and saddle deformations
     Phys. Rev. 126, 220 (1962)

70. J. R. Huizenga, K. M. Clarke, J. E. Gindler, and R. Vandenbosch
     Photofission cross sections of several nuclei with monoenergetic gamma rays

Nucl. Phys. 34, 439 (1962)

71. J. R. Huizenga and R. Vandenbosch
Photoexcitation of the isomeric states of ^{111}Cd and ^{115}In with monoenergetic gamma rays
Nucl. Phys. 34, 457 (1962)

72. R. Vandenbosch and J. R. Huizenga
Kinetic energy distributions of fragments from the fission of Au, Tl, Pb, and Bi
Phys. Rev. 127, 212 (1962)

73. J. Wing and J. R. Huizenga
(p, n) Cross sections of ^{51}V, ^{52}Cr, ^{63}Cu, ^{65}Cu, ^{107}Ag, ^{109}Ag, ^{114}Cd, and ^{139}La from 5 to 10.5 MeV
Phys. Rev. 128, 280 (1962)

74. J. R. Huizenga and R. Vandenbosch
Nuclear fission
Nuclear Reactions, Vol. II, Chapter II, pp. 42–112, P.B. Endt, and P.B. Smith, editors (North-Holland Publishing Company, Amsterdam, The Netherlands, 1962)

75. George L. Bate and J. R. Huizenga
Abundances of ruthenium, osmium and uranium in some cosmic and terrestrial sources
Geochim. Et. Cosmochim. Acta 27, 345 (1963)

76. George L. Bate, R. Chaudhry, and J. R. Huizenga
Fission fragment angular distributions and cross sections for deuteron-induced fission
Phys. Rev. 131, 722 (1963)

77. G. L. Bate and J. R. Huizenga
Fission and reaction cross sections in the uranium region with 4- to 12-MeV protons
Phys. Rev. 133, B1417–B1476 (April 1964)

78. J. P. Unik and J. R. Huizenga
Binary fission studies of the helium-ion induced fission of ^{209}Bi, ^{226}Ra, and ^{238}U
Phys. Rev. 134, B90–B99 (April 1964)

79. J. P. Unik, D. J. Henderson, and J. R. Huizenga
Radioactive species produced by cosmic rays in the Bogou iron meteorite
Geochim. Et Cosmochim, Acta 28, 593 (May 1964)

80. C. T. Bishop, J. R. Huizenga, and J. P. Hummel
Isomer ratios from (α, xn) reactions on silver
Phys. Rev. 135, B401–B411 (July 1964)

81. H. K. Vonach, A. A. Katsanos, and J. R. Huizenga
Cross section fluctuations in the ^{55}Mn (p, α) ^{52}Cr reaction
Phys. Rev. Lett. 13, 88 (July 1964)

82.   K. F. Flynn, L. E. Glendenin, and J. R. Huizenga
Anisotropy of selected fission fragments for helium-ion-induced fission of ^{206}Pb and ^{209}Bi
Nucl. Phys. 58, 321 (September 1964)

83.   B. D. Wilkins, J. P. Unik, and J. R. Huizenga
Studies of nuclear states at the fission saddle-point by inelastic alpha particle scattering
Phys. Lett. 12, 243 (October 1964)

84.   H. K. Vonach, R. Vandenbosch, and J. R. Huizenga
Interpretation of isomer ratios in nuclear reactions with fermi-gas and superconductor models
Nucl. Phys. 60, 70 (1964)

85.   C. T. Bishop, H. K. Vonach, and J. R. Huizenga
Isomer rations for some (n, Γ) reactions
Nucl. Phys. 60, 241 (1964)

86.   J. E. Gindler, G. L. Bate, and J. R. Huizenga
Fission fragment angular distributions in charged-particle-induced fission of ^{226}Ra
Phys. Rev. 136, B1333–B1344 (December 1964)

87.   J. R. Huizenga
Structure of the transition state nucleus in nuclear fission
Nuclear Structure and Electromagnetic Interactions, pp. 319–374, N. MacDonald, editor (Oliver and Boyd, London, 1964)

88.   H. K. Vonach and J. R. Huizenga
Determination of nuclear level densities at an excitation energy of 20 MeV
Phys. Rev. 138, B1372–B1377 (June 1965)

89.   J. R. Huizenga, J. P. Unik, and B. D. Wilkins
On the role of the transition state in nuclear fission
Proc. of IAEA Symp. on Physics and Chemistry of Fission, Salzburg, Austria, Vol. I, pp. 11–22 (1965)

90.   R. Vandenbosch, J. P. Unik, and J. R. Huizenga
Fission-fragment angular, energy, and mass division correlations for the ^{234}U (d, pf) reaction
Proc. of IAEA Symp. on Physics and Chemistry of Fission, Salzburg, Austria, Vol. I, pp. 547–560 (1965)

91.   A. A. Katsanos, J. R. Huizenga, and H. K. Vonach
Nuclear energy levels of ^{56}Fe and ^{59}Co
Phys. Rev. 141, 1053 (1966)

92.   R.F. Reising, G. L. Bate, and J. R. Huizenga
Deformation of the transition state nucleus in energetic fission
Phys. Rev. 141, 1161 (1966)

93.   Curtis R. Keedy, Larry Haskin, James Wing, and J. R. Huizenga

Isomer ratios for the ^{41}K(α,n) $^{44,\ 44M}$Sc, and ^{55}Mn (α,n)$^{58,\ 58M}$ Co reactions
Nucl. Phys. 82, 1 (1966)

94.   H. K. Vonach and J. R. Huizenga
The Co^{59}(p,α)^{56}Fe and ^{56}Fe(p,p^1) reactions
Phys. Rev. 149, 844 (1966)

95.   G. Friedlander, J. O. Rasmussen, J. R. Huizenga, T. T. Sugihara, and A. Turkevich
Nuclear Chemistry, A Current Review, National Academy of Sciences— (National Research Council Publication, 1292C, Washington, DC, 1966)

96.   J. R. Huizenga
The transition state in nuclear fission
Proc. Int. Nucl. Phys. Conf., Gatlinburg, Tenn., pp. 721–736 (Academic Press, 1967)

97.   A. A. Katsanos and J. R. Huizenga
Nuclear energy levels of ^{52}Cr, ^{55}Mn and ^{66}Zn
Phys. Rev. 159, 931 (1967)

98.   J. R. Huizenga and A. A. Katsanos
Distribution of spacings of nuclear energy levels of mixed spin and parity
Nucl. Phys. A. 98, 614 (1967)

99.   W. Loveland, J. R. Huizenga, A. Behkami, and J. H. Roberts
Identification of the single-particle states in the fission transition nucleus ^{235}U
Phys. Lett. 24B, 666 (1967)

100.   R. Vandenbosch, K. L. Wolf, J. Unik, C. Stephan, and J. R. Huizenga
Dependence of fission fragment angular distributions on target deformation and spin
Phys. Rev. Lett. 19, 1138 (1967)

101.   Th. W. Elze, T. von Egidy, and J. R. Huizenga
(^{3}He,α) Reactions on actinide elements
Phys. Lett. 27B, 78 (1968)

102.   A. N. Behkami, J. H. Roberts, W. Loveland, and J. R. Huizenga
Structure of the fission transition nucleus ^{235}U
Phys. Rev. 171, 1267 (1968)

103.   J. R. Huizenga, A. N. Behkami, J. W. Meadows Jr., and E. D. Klema
Fragment angular distributions for mono-energetic neutron-induced fission of ^{239}Pu
Proc. Conf. on Neutron Cross Sections and Technology at Washington, DC, Vol. 1, p. 603 (1968)

104.   A. N. Behkami, J. R. Huizenga, and J. H. Roberts
Angular distributions and cross sections of fragments from neutron-induced fission of ^{232}Th
Nucl. Phys. A 118, 65 (1968)

105. J. E. Gindler and J. R. Huizenga
Nuclear fission
In *Treatise on Nuclear Chemistry,* Vol. II, pp. 1–184 (Academic Press, New York and London, 1968)

106. J. R. Huizenga, A. N. Behkami, J. W. Meadows, and E. D. Klema
Nuclear pairing energy of transition nucleus ^{240}Pu
Phys. Rev. 174, 1539 (1968)

107. H. K. Vonach, A. A. Katsanos, and J. R. Huizenga
Determination of the level width and density of ^{32}S between 17 and 21 MeV excitation energy
Nucl. Phys. A 122, 465 (1968)

108. J. R. Huizenga, A. N. Behkami, and L. G. Moretto
Note on interpretation of fission-fragment angular distributions at moderate excitation energies
Phys. Rev. 177, 1826 (1969)

109. L. G. Moretto, R. C. Gatti, S. G. Thompson, J. R. Huizenga, and J. O. Rasmussen
Pairing effects at the fission saddle point of ^{210}Po and ^{211}Po
Phys. Rev. 178, 1845 (1969)

110. Th. W. Elze, T. v. Egidy, and J. R. Huizenga
A study of the ^{232}Th (^{3}He, α) ^{231}Th reaction
Nucl. Phys. A 128, 564 (1969)

111. Th. W. Elze and J. R. Huizenga
A study of the ^{236}U (^{3}He, α) ^{235}U reaction
Nucl. Phys. A 133, 10 (1969)

112. J. R. Huizenga, H. K. Vonach, A. A. Katsanos, A. J. Gorski, and C. J. Stephan
Level densities from excitation functions of isolated levels
Phys. Rev. 182, 1149 (1969)

113. J. R. Huizenga, A. N. Behkami, and J. H. Roberts
Channel analysis of neutron-induced fission of ^{236}U
Proc. Second Symposium on Physics and Chemistry of Fission, IAEA, Vienna, 403 (1969)

114. Th. W. Elze and J. R. Huizenga
Levels of ^{237}Np excited by the ^{236}U (^{3}He,d) ^{237}Np and ^{236}U (α,t) ^{237}Np reactions
Phys. Rev. 1C, 328 (1970)

115. K. Otozai, J. W. Meadows, A. N. Behkami, and J. R. Huizenga
Fragment angular distributions from neutron-induced fission of ^{242}Pu
Nucl. Phys. A 144, 502 (1970)

116. T. von Egidy, Th. W. Elze, and J. R. Huizenga
Nuclear levels of ^{237}U excited by the (^{3}He, α) reaction
Nucl. Phys. A 145, 306 (1970)

117. J. R. Huizenga
Nuclear fission revisited
Science 168, 1405 (June 19, 1970)

118. Th. W. Elze and J. R. Huizenga
Energy levels of ^{243}Am excited by the ^{242}Pu (^{3}He, d) and ^{242}Pu(α,t) reactions
Nucl. Phys. A 149, 585 (1970)

119. J. S. Boyno, T.W. Elze, and J. R. Huizenga
A study of the ^{232}Th (d,t) ^{231}Th and ^{238}U (d, t) ^{237}U reactions
Phys. Rev. A 157, 263 (1970)

120. Th. W. Elze and J. R. Huizenga
A study of the ^{242}Pu (d,t) ^{241}Pu and ^{242}Pu (^{3}He, α) ^{241}Pu reactions
Phys. Rev. C 3, 234 (1971)

121. C. C. Lu, J. R. Huizenga, C. J. Stephan, and A. J. Gorski
Effects of isospin on statistical nuclear decay
Nucl. Phys. A 164, 225 (1971)

122. G. Yuen, G. T. Rizzo, A. N. Behkami, and J. R. Huizenga
Fragment angular distributions of neutron-induced fission of ^{230}Th
Nucl. Phys. A 171, 614 (1971)

123. L. C. Vaz, C. C. Lu, and J. R. Huizenga
Isospin and the Bohr independence hypothesis
Phys. Rev. C 5, 463 (1972)

124. J. R. Huizenga
Experimental and theoretical nuclear level densities
Statistical Properties of Nuclei, p. 425, J. B. Garg, editor (Plenum Press, New York, 1972)

125. J. R. Huizenga
Near barrier fission induced with photons
Nucl. Technology 13, 20 (1972)

126. F. C. Williams Jr., G. Chan, and J. R. Huizenga
The significance of shell corrections in the parameterization of numerical state density calculations
Nucl. Phys. A 187, 225 (1972)

127. Th. W. Elze, J. S. Boyno, and J. R. Huizenga
Collective states of Gd isotopes from (p,t) reactions
Nucl. Phys. A 187, 473 (1972)

128. Th. W. Elze and J. R. Huizenga
Collective states of ^{232}Th, ^{238}U and ^{242}Pu
Nucl. Phys. A 187, 545 (1972)

129. C. C. Lu, L. C. Vaz, and J. R. Huizenga
Level densities from spectra of different reactions populating the same nucleus
Nucl. Phys. A 190, 229 (1972)

130. J. S. Boyno and J. R. Huizenga
Levels of [155, 157, 159, 161]Tb excited in helium-induced single-proton-transfer reactions
Phys. Rev. C 6, 144 (1972)

131. L. C. Vaz, C. C. Lu, and J. R. Huizenga
Level densities from analysis of particle spectra
Phys. Rev. C 6, 1896 (1972)

132. C. C. Lu, L. C. Vaz, and J. R. Huizenga
Spin distribution of the nuclear level density near A=60
Nucl. Phys. A 197, 321 (1972)

133. J. R. Huizenga and L. G. Moretto
Nuclear level densities
Ann. Rev. Nucl. Sci. 22, 427 (1972)

134. J. R. Huizenga and H. C. Britt
Threshold photofission: theory and experiment
Proc. Int. Conf. on Photonuclear Reactions and Applications V2, 833 (1973)

135. J. S. Boyno, J. R. Huizenga, Th. W. Elze, and C.E. Bemis Jr.
Levels of ^{234}U and ^{236}U excited by the (d,d') reaction
Nucl. Phys. A 209, 125 (1973)

136. W. K. Hensley, W. A. Bassett, and J. R. Huizenga
Pressure dependence of the radioactive decay constant of beryllium-7
Science 181, 1164 (1973)

137. M. Kildir and J. R. Huizenga
Isospin dependence of the nuclear level width
Phys. Rev. C 8, 1965 (1973)

138. R. Vandenbosch and J. R. Huizenga
Nuclear Fission (Academic Press, New York, 1973)

139. A. N. Behkami and J. R. Huizenga
Comparison of experimental level densities and spin cutoff factors with microscopic theory for nuclei near A=60
Nucl. Phys. A 217 78 (1973)

140. H. Freiesleben, H. C. Britt, and J. R. Huizenga
Energy dependence of Γ_f / Γ_n for the nucleus ^{216}Rn
Proc. Third Int. Symposium on Physics and Chemistry of Fission, IAEA, Vienna
1, 447 (1974)

141. H. C. Britt and J. R. Huizenga
Reevaluation of experimental estimates of the pairing gap at the fission saddle point
Phys. Rev. C 9 435 (1974)

142. C. Kalbach-Cline, J. R. Huizenga, and H. K. Vonach
Isospin conservation and preequilibrium decay in (p, p^1) reactions on neutron rich tin isotopes
Nucl. Phys. A 222, 405 (1974)

143. J. R. Huizenga, A. N. Behkami, J. S. Sventek, and R.W. Atcher
Comparison of neutron resonance spacings with microscopic theory for spherical nuclei
Nucl, Phys. A 223, 577 (1974)

144. J. R. Huizenga, A. N. Behkami, R.W. Atcher, J. S. Sventek, H. C. Britt, and H. Freiesleben
Comparison of neutron resonance spacings with microscopic theory for nuclei with static deformation
Nucl. Phys. A 223, 589 (1974)

145. H. Freiesleben and J. R. Huizenga
Total reaction cross sections of deformed nuclei: a study of the $^{233,\,238}$U+α systems
Nucl. Phys. A 224, 503 (1974)

146. H. Freiesleben, H. C. Britt, J. Birkelund, and J. R. Huizenga
^{6}Li, ^{7}Li induced reactions on ^{209}Bi
Phys. Rev. 10C, 245 (1974)

147. K. L. Wolf, J. P. Unik, J. R. Huizenga, J. Birkelund, H. Freiesleben, and V. E. Viola
A study of strongly damped collisions in the reaction of 600-MeV ^{84}Kr on a ^{209}Bi target
Phys. Rev. Lett. 33, 1105 (1974)

148. T. von Egidy, Th. W. Elze, and J. R. Huizenga
Nuclear levels of ^{239}Np excited by the (^{3}He, d) and (α, t) reactions
Phys. Rev. C 11, 529 (1975)

149. Th. W. Elze and J. R. Huizenga
Levels of ^{233}Pa excited in helium-induced single-proton transfer reactions
Z. Physik A272, 119 (1975)

150. J. P. Bondorf, J. R. Huizenga, M.I. Sobel, and D. Sperber
A classical model for strongly damped collisions in heavy ion reactions
Phys. Rev. C 11, 1265 (1975)

151. J. R. Huizenga
Selected aspects of very heavy ion reactions
Nukleonika 20, 291–344 (1975)

152. R. C. Thompson, J. S. Boyno, J. R. Huizenga, D. G. Burke, and Th. W. Elze
A study of the ^{188}Os (p,t) ^{186}Os, ^{190}Os (p,t) ^{188}Os and ^{192}Os (p,t) ^{190}Os reactions
Nucl. Phys. A 242, 1 (1975)

153. H. Freiesleben, G. T. Rizzo, and J. R. Huizenga
^{6}Li and ^{7}Li induced fission of ^{232}Th and ^{238}U
Phys. Rev. C 12, 42 (1975)

154. R. C. Thompson, J. R. Huizenga, and Th. W. Elze
     Collective states in ^{230}Th, ^{240}Pu, ^{244}Pu and ^{248}Cm excited by inelastic deuteron scattering
     Phys. Rev. C 12, 1227 (1975)

155. J. R. Birkelund, J. R. Huizenga, H. Freiesleben, K. L. Wolf, J. P. Unik, and V. E. Viola
     The elastic scattering of ^{40}Ar and ^{84}Kr on ^{209}Bi and ^{238}U
     Phys. Rev. C 13, 133 (1976)

156. R. C. Thompson, J. R. Huizenga, and Th. W. Elze
     Collective states in ^{233}U, ^{235}U, ^{237}Np, and ^{239}Pu excited by inelastic deuteron scattering
     Phys. Rev. C 13, 638 (1976)

157. W. U. Schröder, J. R. Birkelund, J. R. Huizenga, K. L. Wolf, J. P. Unik, and V. E. Viola
     Study of the ^{209}Bi + ^{136}Xe reaction
     Phys. Rev. Lett. 36, 514 (1976)

158. J. R. Huizenga
     Nuclear reactions revisited with very heavy ions
     Accounts of Chemical Research 9, 325 (1976)

159. J. R. Huizenga, J. R. Birkelund, and M. W. Johnson
     Elastic scattering of very heavy ions
     Proc. Symposium on Macroscopic Features of Heavy-Ion Collisions, Argonne National Laboratory, Report ANL/PHY-76–2, *I* 1–32 (1976)

160. J. R. Birkelund, W. U. Schröder, J. R. Huizenga, K. L. Wolf, J. P. Unik, and V. E. Viola
     Fragment charge distributions from the ^{209}Bi and ^{136}Xe reaction at 1130 MEV
     Proc. Symposium on Macroscopic Features of Heavy-Ion Collisions, Argonne National Laboratory, Report ANL/PHY-76–2, II, 451–462 (1976)

161. J. R. Huizenga, J. R. Birkelund, W. U. Schröder, K. L. Wolf, and V. E. Viola
     Energy dissipation and nucleon transfer in heavy ion reactions
     Phys. Rev. Lett. 37, 885 (1976)

162. J. R. Huizenga
     A new reaction process: strongly damped or deep inelastic collisions
     Comments on Nucl. and Particle Phys. 7, 17 (1977)

163. J. R. Huizenga
     Nuclear fission
     McGraw-Hill Encyclopedia of Science and Technology, Vol. 5, 303 (1977)

164. R. C. Thompson, W. W. Wilcke, J. R. Huizenga, W. K. Hensley, and D. G. Perry
     Levels of isotopes $^{233, 235, 237}$Pa and $^{229, 231}$Ac studied by the (t, α) reaction
     Phys. Rev. C 15, 2019 (1977)

165. M. W. Johnson, W. U. Schröder, J. R. Huizenga, W. K. Hensley, D. G. Perry, and J.C. Browne
Measurement of total muon-capture rates in ^{232}Th, $^{235,\,238}$U, and ^{239}Pu
Phys. Rev. C 15, 2169 (1977)

166. W. U. Schröder, J. R. Birkelund, J. R. Huizenga, K. L. Wolf, and V. E. Viola Jr.
Interaction times for damped heavy-ion collisions
Phys. Rev. C 16, 623 (1977)

167. W. W. Wilcke, W. Feix, Th. W. Elze, J. R. Huizenga, R. C. Thompson, and R.M. Dreizler
Influence of form factors and multistep effects on the ^{232}Th(d, t) ^{231}Th reaction
Nucl. Phys. A 286, 297 (1977)

168. W. Feix, W. W. Wilcke, Th. W. Elze, H. Rebel, J. R. Huizenga, R. C. Thompson, and R.M. Dreizler
Study of the nuclear-coulomb interference effects in inelastic deuteron scattering on ^{238}U
Phys. Lett. 69B, 407 (1977)

169. J. R. Huizenga, W. U. Schröder, J. R. Birkelund, and W. W. Wilcke
Energy dissipation, mass diffusion and interaction times for heavy-ion collisions
Proc. Meeting on Heavy-Ion Collisions, Fall Creek Falls State Park, Pikeville, Tenn., CONF-770602, p. 179 (1977)

170. W. U. Schröder, J. R. Huizenga, J. R. Birkelund, K. L. Wolf, and V. E. Viola
Dissipation, mass exchange, and the microscopic time scale of heavy-ion collisions
Phys. Lett. 71B, 283 (1977)

171. J. R. Birkelund and J. R. Huizenga
Internuclear potentials from heavy-ion fusion excitation functions
Proc. Symposium on Heavy-Ion Elastic Scattering, University of Rochester, October 25–26 (1977)

172. W. U. Schröder and J. R. Huizenga
Damped heavy-ion collisions
Ann. Rev. Nucl. Sci. 27, 465–547 (1977)

173. J. R. Birkelund and J. R. Huizenga
The determination of internuclear potentials from heavy-ion fusion excitation functions
Phys. Rev. C 17, 126 (1978)

174. W. U. Schröder, J. R. Huizenga, J. R. Birkelund, and W. W. Wilcke
The mechanisms of energy dissipation and nucleon exchange in damped reactions induced by very heavy ions
Proc. Workshop on Gross Properties of Nuclei and Nuclear Excitations VI, Hirschegg, Austria, p. 85, January (1978)

175. J. R. Huizenga
     Nuclear properties of einsteinium and fermium
     Proc. Symposium Commemorating the 25th Anniversary of the Discovery of Elements 99 and 100, Lawrence Berkeley Laboratory, Jan. 23, 1978, LBL-7701, pp. 17–26

176. M. W. Johnson, R. C. Thompson, and J. R. Huizenga
     States in ^{233}U excited by the ^{234}U (d,t) and ^{234}U (3He, α) reactions
     Phys. Rev. C 17, 927 (1978)

177. J. R. Huizenga
     Nuclear reactions
     McGraw-Hill Yearbook of Science and Technology, p. 276 (1978)

178. J. R. Birkelund, J. R. Huizenga, J. N. De, and D. Sperber
     Heavy-ion fusion based on the proximity potential and one-body friction
     Phys. Rev. Lett. 40, 1123 (1978)

179. R. D. Griffioen, R. C. Thompson, and J. R. Huizenga
     Level of ^{235}Np excited by the ^{234}U(3He, d) and ^{234}U(α, t) reactions
     Phys. Rev. C 18, 671 (1978)

180. W. U. Schröder, J. R. Birkelund, J. R. Huizenga, K. L. Wolf, and V. E. Viola Jr.
     Mechanisms of very heavy ion collisions; the ^{209}Bi + ^{136}Xe reaction at E_{LAB} = 1130 MEV
     Phys. Reports 45, No. 5, 301 (1978)

181. W. W. Wilcke, M. W. Johnson, W. U. Schröder, J. R. Huizenga, and D. G. Perry
     Neutron emission from actinide muonic atoms
     Phys. Rev. C 18, 1452 (1978)

182. S.L. Tabor, D. A. Goldberg, and J. R. Huizenga
     Relevance of the proximity potential to light ion-scattering
     Phys. Rev. Lett. 41, 1285 (1978)

183. J. N. De, A. Sherman, D. Sperber, J. R. Birkelund, and J. R. Huizenga
     Fusion-excitation functions as a test of the radial dependence of the proximity potential
     S. Afr. J. Phys. 1, No. 3/4, 239 (1978)

184. J. R. Huizenga, J. R. Birkelund, L. E. Tubbs, J. N. De, and D. Sperber
     Role of conservative potential and dissipation in heavy-ion fusion
     Proc. Int. Workshop on Gross Properties of Nuclei and Nuclear Excitations VII, Hirschegg, Austria, p. 45, INKA-Conf-79–001–017, January (1979)

185. D. Hilscher, W. W. Wilcke, W. U. Schröder, A. D. Hoover, J. R. Birkelund, J. R. Huizenga, A. Mignerey, K. L. Wolf, H. Breuer, and V. E. Viola
     Neutron emission from strongly damped reaction ^{165}Ho + ^{56}Fe at 8.5 MeV/u
     Proc. Int. Workshop on Gross Properties of Nuclei and Nuclear Excitations VII, Hirschegg, Austria, p. 94, INKA-Conf-79–001–017, January (1979)

186. H. Breuer, B. G. Glagola, V. E. Viola, K. L. Wolf, A. C. Mignerey, J. R. Birkelund, D. Hilscher, A. D. Hoover, J. R. Huizenga, W. U. Schröder, and W. W. Wilcke
Nucleon exchange and A/Z equilibrium in interactions of 8.3 MeV/u ^{56}Fe ions with ^{56}Fe, ^{165}Ho and ^{209}Bi
Phys. Rev. Lett. 43, 191 (1979)

187. D. Hilscher, J. R. Birkelund, A. D. Hoover, W. U. Schröder, W. W. Wilcke, J. R. Huizenga, A. Mignerey, K. L. Wolf, H. F. Breuer, and V. E. Viola Jr.
Structure in the energy spectra of the fragments produced in damped heavy-ion reactions
Phys. Rev. C 20, 556 (1979)

188. D. Hilscher, J. R. Birkelund, A. D. Hoover, W. U. Schröder, W. W. Wilcke, J. R. Huizenga, A. Mignerey, K. L. Wolf, H. Breuer, and V. E. Viola Jr.
Neutron emission from the ^{165}Ho + ^{56}Fe reaction at 8.5 MeV/u
Phys. Rev. C. 20, 576 (1979)

189. J. R. Huizenga, J. R. Birkelund, L. E. Tubbs, J. N. De, and D. Sperber
Fusion reactions at high energies
Proc. Symposium on Heavy Ion Physics from 10 to 200 MeV/amu, Brookhaven National Laboratory, Report BNL-51115, Vol. 1, pp. 235–271 (1979)

190. W. U. Schröder, W. W. Wilcke, M. W. Johnson, D. Hilscher, J. R. Huizenga, J.C. Browne, and D. G. Perry
Evidence for atomic muon capture by fragments from prompt fission of muonic ^{237}Np, ^{239}Pu, and ^{242}Pu
Phys. Rev. Lett. 43, 672 (1979)

191. J. R. Birkelund, L. E. Tubbs, J. R. Huizenga, J.M. De, and D. Sperber
Heavy-ion fusion: comparison of experimental data with classical trajectory models
Phys. Reports 56, No. 3, 107 (1979)

192. J. R. Huizenga
Heavy-ion reactions: a new frontier of nuclear science
Proc. Symposium on Deep-Inelastic and Fusion Reactions with Heavy Ions, Berlin. Lecture notes in Physics 117, 1–24, W. von Oertzen, editor (Springer-Verlag, 1979)

193. D. Hilscher, J. R. Birkelund, A. D. Hoover, W. U. Schröder, W. W. Wilcke, J. R. Huizenga, A. C. Mignerey, K. L. Wolf, H. F. Breuer, and V. E. Viola
Neutron emission in heavy ion reactions
Proc. Symposium on Deep-Inelastic and Fusion Reactions with Heavy Ions, Berlin. Lecture Notes in Physics 117, 100–112, W. von Oertzen, editor (Springer-Verlag, 1979)

194. J. R. Birkelund, L. E. Tubbs, J. R. Huizenga, J. N. De, and D. Sperber
Heavy-ion fusion: a classical trajectory model
Proc. Symposium on Deep-Inelastic and Fusion Reactions with Heavy Ions, Berlin. Lecture Notes in Physics 117, 294–311, W. von Oertzen, editor (Springer-Verlag, 1979)

195. W. U. Schröder, J. R. Birkelund, J. R. Huizenga, W. W. Wilcke, and J. Randrup
     The effect of Pauli blocking on exchange and dissipation mechanisms operating in heavy ion reactions
     Phys. Rev. Lett. 44, 308 (1980)

196. J. R. Huizenga
     Experimental summary of the international symposium on continuum spectra of heavy ion reactions
     Nuclear Science Research Conf. Series, Vol. *2*, 405–443, T. Tamura, J. B. Natowitz, and D. H. Youngblood, editors (Harwood Academic Publishers, 1980)

197. W. U. Schröder, J. R. Birkelund, J. R. Huizenga, W. W. Wilcke, and J. Randrup
     Manifestation of the quantum-statistical nature of exchange and dissipation mechanisms operating in damped nuclear reactions
     Proc. Int. Workshop on Gross Properties of Nuclei and Nuclear Excitations VIII, Hirschegg, Austria, p. 92, January 14–19 (1980)

198. W. W. Wilcke, M. W. Johnson, W. U. Schröder, D. Hilscher, J. R. Birkelund, J. R. Huizenga, J.C. Browne, and D. G. Perry
     Actinide muonic atom lifetimes deduced from muon-induced fission
     Phys. Rev. C 21, 2019 (1980)

199. R.W. Atcher, A. M. Friedman, J. R. Huizenga, G.V.S. Rayudu, E.A. Silverstein, and D. A. Turner
     Manganese-52m, a new short-lived, generator-produced radio-nuclide: a potential generator for positron tomography
     Journal Nucl. Medicine 21, 565 (1980)

200. R.W. Atcher, A. M. Friedman, and J. R. Huizenga
     Production of ^{52}Fe for use in a radionuclide generator system
     Int. Journal Nucl. Medicine and Biology 7, 75 (1980)

201. V. E. Viola, A. C. Mignerey, H. Breuer, K. L. Wolf, B. G. Glagola, W. W. Wilcke, W. U. Schröder, J. R. Huizenga, D. Hilscher, and J. R. Birkelund
     Possible production of actinide spontaneous fission activities in damped collisions of ^{209}Bi + ^{56}Fe
     Phys. Rev. C 22, 122 (1980)

202. W. W. Wilcke, J. R. Birkelund, A. D. Hoover, J. R. Huizenga, W. U. Schröder, V. E. Viola, K. L. Wolf, and A. C. Mignerey
     Bombarding energy dependence of the ^{209}Bi + ^{136}Xe reaction
     Phys. Rev C 22, 128 (1980)

203. A. C. Mignerey, V. E. Viola, H. Breuer, K. L. Wolf, B. G. Glagola, J. R. Birkelund, D. Hilscher, J. R. Huizenga, W. U. Schröder, and W. W. Wilcke
     Dependence of isobaric charge distributions on energy loss and mass asymmetry in damped collisions
     Phys. Rev. Lett. 45, 509 (1980)

204. H. Breuer, K. L. Wolf, B. G. Glagola, K.K. Kwiatkowski, A. C. Mignerey, V. E. Viola, W. W. Wilcke, W. U. Schröder, J. R. Huizenga, D. Hilscher, and J. R. Birkelund
Production of neutron-excess nuclei in ^{56}Fe-induced reactions
Phys. Rev. C 22, 2454 (1980)

205. W. W. Wilcke, J. R. Birkelund, H. J. Wollersheim, A. D. Hoover, J. R. Huizenga, W. U. Schröder, and L. E. Tubbs
Reaction parameters for heavy-ion collisions
Atomic Data and Nuclear Data Tables 25, 389–619 (1980)

206. W. U. Schröder, J. R. Huizenga, and J. Randrup
Correlated mass and charge transport induced by statistical nucleon exchange in damped nuclear reactions
Phys. Lett. 98B, 355 (1981)

207. W. U. Schröder and J. R. Huizenga
Dissipative reaction dynamics of heavy-ion collisions
Comments on Nucl. and Particle Phys. 10, 19 (1981)

208. W. W. Wilcke, J. R. Birkelund, J. P. Kosky, H. J. Wollersheim, A. D. Hoover, J. R. Huizenga, W. U. Schröder, L. E. Tubbs, and D. Hilscer
A two dimensional position sensitive ΔE-E counter for energetic light charged particles
Nucl. Instr. and Methods 188, 293 (1981)

209. J. R. Huizenga, J. R. Birkelund, W. U. Schröder, W. W. Wilcke, and H. J. Wollersheim
Heavy-ion fusion revisited
Dynamics of Heavy-Ion Collisions, pp. 15–39, N. Cindro, R.A. Ricci, and W. Greiner, editors (North-Holland Publishing Company, Amsterdam, 1981)

210. H. J. Wollersheim, W. W. Wilcke, J. R. Birkelund, J. R. Huizenga, W. U. Schröder, H. Freiesleben, and D. Hilscher
The ^{209}Bi + ^{136}Xe reaction at E_{lab} = 1422 MeV
Phys. Rev C 24, 2114 (1981)

211. J. R. Huizenga, J. R. Birkelund, W. U. Schröder, W. W. Wilcke, and H. J. Wollersheim
Experimental characteristics of damped nuclear reactions
Nuclear Physics (Proc. Nuclear Physics Workshop, I.C.T.P., Trieste), pp. 583–600, C.H. Dasso, editor (North-Holland Publishing Company, Amsterdam, 1981)

212. A. D. Hoover, J. R. Birkelund, D. Hilscher, W. U. Schröder, W. W. Wilcke, J. R. Huizenga, H. Breuer, A. C. Mignerey, V. E. Viola, and K. L. Wolf
The ^{165}Ho + ^{56}Fe reaction at E_{lab} = 462 MeV
Phys. Rev C 25, 256 (1982)

213. H. J. Wollersheim, W. W. Wilcke, J. R. Birkelund, and J. R. Huizenga
Influence of the impact parameter on element distributions in dissipative heavy-ion collisions
Phys. Rev. C 25, 338 (1982)

214. J. R. Huizenga
      Deep inelastic collisions
      McGraw-Hill Encyclopedia of Science and Technology, 5th edition, p. 63 (1982)
215. J. R. Birkelund, H. Freiesleben, J. R. Huizenga, W. U. Schröder, W. W.
      Wilcke, K. L. Wolf, J. P. Unik, and V. E. Viola Jr.
      Mechanisms of heavy-ion dissipative collisions: the ^{209}Bi + ^{84}Kr reaction at
      E_{lab} = 712 MeV
      Phys. Rev. C 26, 1984 (1982)
216. J. R. Huizenga, W. U. Schröder, J. R. Birkelund, and W. W. Wilcke
      Distinctive features of heavy ion dissipative collisions
      Nucl. Phys. A 387, 257C (1982)
217. J. R. Birkelund, A. D. Hoover, J. R. Huizenga, W. U. Schröder, and W. W.
      Wilcke
      Fusion cross sections for the ^{165}Ho + ^{56}Fe System
      Phys. Rev. C 27, 882 (1983)
218. H. H. Rossner, D. Hilscher, E. Holub, G. Ingold, U. Jahnke, H. Orf, J. R.
      Huizenga, J. R. Birkelund, W. U. Schröder, and W. W. Wilcke
      Angular distributions of fragments from fission induced by 220-MeV ^{20}Ne
      on targets of ^{165}Ho, ^{197}Au, and ^{209}Bi
      Phys. Rev. C 27, 2666 (1983)
219. W. W. Wilcke, J. P. Kosky, J. R. Birkelund, M. A. Butler, A. D. Dougan, J. R.
      Huizenga, W. U. Schröder, and H. J. Wollersheim
      A new mechanism of α-particle production in heavy-ion induced fission
      Phys. Rev. Lett. 51, 99 (1983)
220. H. Breuer, A. C. Mignerey, V. E. Viola, K. L. Wolf, J. R. Birkelund, D.
      Hilscher, J. R. Huizenga, W. U. Schröder, and W. W. Wilcke
      Charge and mass exchange in ^{56}Fe-induced reactions at 8.3 MeV/nucleon
      Phys. Rev. C 28, 1080 (1983)
221. J. R. Huizenga, J. R. Birkelund, L. E. Tubbs, D. Hilscher, U. Jahnke, H. H.
      Rossner, B. Gebauer, and H. Lettau
      Limitation of complete linear momentum transfer in heavy-ion reactions
      Phys. Rev. C 28, 1853 (1983)
222. J. P. Kosky, W. W. Wilcke, J. R. Birkelund, M. A. Butler, A. D. Dougan, J. R.
      Huizenga, W. U. Schröder, H. J. Wollersheim, and D. Hilscher
      Alpha-particle emission from the strongly damped reaction of ^{165}Ho +
      ^{56}Fe at 8.3 MeV/nucleon
      Phys. Lett. 133B, 153 (1983)
223. J. R. Birkelund and J. R. Huizenga
      Fusion reactions between heavy nuclei
      Ann. Rev. Nucl. and Part. Sci. 33, 265–322 (1983)
224. J. R. Huizenga
      Deep inelastic reactions and fusion
      Nucl. Phys. A 409, 181c (1983)

225. A. C. Mignerey, K. L. Wolf, D. G. Raich, V. E. Viola, J. R. Birkelund, W. U. Schröder, and J. R. Huizenga
Bombarding energy dependence of the ^{144}Sm + ^{84}Kr reaction
Phys. Rev. C 29, 158–182 (1984)

226. H. H. Rossner, J. R. Huizenga, and W. U. Schröder
Statistical scission model of fission fragment angular distributions
Phys. Rev. Lett. 53, 38 (1984)

227. J. R. Birkelund and J. R. Huizenga
Measurement of nuclear potentials from fusion excitation functions
Phys. Rev. C 30, 401 (1984)

228. W. U. Schröder and J. R. Huizenga
Damped nuclear reactions
Treatise on Heavy-Ion Science, Vol. 2, pp. 113–726, D. Allan Bromley, editor
(Plenum Press, 1984)

229. J. R. Huizenga and W. U. Schröder
Nuclear transport phenomena in low-energy heavy-ion collisions
Semiclassical Descriptions of Atomic and Nuclear Collisions (Proc. Niels Bohr Centennial Conf., Copenhagen, March 25–28, 1985), pp. 255–280, J. Bang and J. De Boer, editors (North Holland, Amsterdam, 1985)

230. L. E. Tubbs, J. R. Birkelund, J. R. Huizenga, D. Hilscher, U. Jahnke, H. H. Rossner, and B. Gebauer
Linear momentum transfer in 292-MEV ^{20}Ne-induced fission of ^{165}Ho, ^{181}Ta, ^{197}Au, ^{209}Bi, and ^{238}U
Phys. Rev. C 32, 214 (1985)

231. J. R. Huizenga, H. H. Rossner, and W. U. Schröder
Fragment angular distributions from heavy-ion induced fission
Proc. 1984 INS-RIKEN Int. Symp. Heavy Ion Physics, J. Phys. Soc. Japan 54, Suppl. II, pp. 257–271 (1985)

232. H. H. Rossner, J. R. Huizenga, and W. U. Schröder
Fission fragment angular distributions
Phys. Rev. C 33, 560 (1986)

233. W. U. Schröder, J. R. Birkelund, J. R. Huizenga, L. E. Tubbs, W. W. Wilcke, W. Bohne, B. Gebauer, D. Hilscher, E. Holub, G. Ingold, U. Jahnke, H. Lettau, H. Morgenstern, H. Orf, H. H. Rossner, W. P. Zank, H. Gemmeke, K. Keller, L. Lassen, and W. Lucking
Dissipation of linear momentum and energy in fusion-like reactions
Proc. Berlin Symposium on Nucl. Physics with Heavy Ions at VICKSI, February (1986)

234. J. Töke, J. R. Birkelund, M. A. Butler, J. R. Huizenga, J. P. Kosky, L. M. Schmieder, W. U. Schröder, and W. W. Wilcke
Signatures of fast alpha-particle emission in damped heavy-ion reactions
Proc. Worksh. Nuclear Dynamics IV, Copper Mountain, Colorado, pp. 66–69, February 24–28 (1986)

235. W. U. Schröder, R. T. de Souza, J. R. Huizenga, and L. M. Schmieder
      Mass, charge, and energy transfer in damped reactions
      *Proc. Int. Symposium on Nuclear Fission and Heavy-Ion-Induced Reactions,
      University of Rochester, Rochester, New York, April 21 and 22, 1986,* Nucl. Res.
      Conf. Series, Vol. 11, pp. 255–291 (Harwood Academic Publishers, New
      York, 1987)

236. H. H. Rossner, J. R. Huizenga, and W. U. Schröder
      The fission channel in heavy-ion reactions
      *Proc. Int. Symposium on Nuclear Fission and Heavy-Ion-Induced Reactions,
      University of Rochester, Rochester, New York, April 21 and 22, 1986,* Nucl. Res.
      Conf. Series, Vol. 11, pp. 127–142 (Harwood Academic Publishers, New
      York, 1987)

237. M. A. Butler, S. S. Datta, R. T. de Souza, J. R. Huizenga, W. U. Schröder, J.
      Töke, and J. L. Wile
      Relaxation of the mass-asymmetry degree of freedom in heavy-ion reactions
      Phys. Rev. C 34, 2016 (1986)

238. J. L. Wile, W. U. Schröder, J. R. Huizenga, and D. Hilscher
      Temperatures, energies, and degree of thermal equilibration of fragments
      in damped nuclear reactions
      Phys. Rev. C 35, 1608 (1987)

239. I. M. Govil, J. R. Huizenga, W. U. Schröder, and J. Töke
      Shapes of alpha-particle spectra from energetic heavy-ion fusion reactions
      Phys. Lett. B 197, 515 (1987)

240. S. S. Datta, J. R. Huizenga, W. U. Schröder, R. T. de Souza, J. Töke, and J.
      L. Wile
      Energy dissipation and particle emission in heavy-ion reactions
      *Proc. Int. Symposium on Dynamics of Collective Phenomena in Nuclear and Sub-
      nuclear Long Range Interactions in Nuclei, Bad Honnef, Germany, May 4–7,
      1987,* pp. 273–287, Peter David, editor (World Scientific Publishing Com-
      pany, 1988)

241. J. R. Huizenga, M. A. Butler, H. H. Rossner, J. L. Wile, S. S. Datta, R. T. de
      Souza, D. Hilscher, W. U. Schröder, and J. Töke
      Massive heavy-ion reactions
      *Proc. All-Union Symposium on the Physics of Nuclear Fission, Obninsk, USSR,
      June 2–5, 1987,* Proceedings in Russian, pp. 65–75 (1988)

242. R. T. de Souza, W. U. Schröder, J. R. Huizenga, R. Planeta, K. Kwiatkowski,
      V. E. Viola, and H. Breuer
      Evolution of mass and charge asymmetry in damped heavy-ion
      Reactions
      Phys. Rev. C 37, 1783 (1988)

243. R. T. de Souza, J. R. Huizenga, and W. U. Schröder
Effect of a steep gradient on the potential energy surface on nucleon exchange
Phys. Rev., C 37, 1901 (1988)

244. R. Planeta, S. H. Zhou, K. Kwiatkowski, W. G. Wilson, V. E. Viola, H. Breuer, D. Benton, F. Khazaie, R. J. McDonald, A. C. Mignerey, A. Weston-Dawkes, R. T. de Souza, J. R. Huizenga, and W. U. Schröder
N/Z equilibration in damped collisions induced by E/A = 8.5 MeV ^{58}Ni and ^{64}Ni on ^{238}U
Phys. Rev., C 38, 195 (1988)

245. W. U. Schröder, S. S. Datta, J. L. Wile, J. Tōke, R. T. de Souza, and J. R. Huizenga
Energy relaxation in damped reactions
Proc. Texas A & M Symp. on Hot Nuclei, pp. 223–240, S. Shlomo, R.P. Schmitt, and J. B. Natowitz, editors (World Scientific Publishing Company, 1988)

246. R. T. de Souza, W. U. Schröder, J. R. Huizenga, J. Tōke, S. S. Datta, and J. L. Wile
Nucleon exchange in the absence of strong driving forces: the reaction ^{238}U + ^{48}Ca at E_{lab} = 425 MeV
Phys. Rev., C39, 114 (1989)

247. J. L. Wile, S. S. Datta, W. U. Schröder, J. R. Huizenga, J. Tōke, and R. T. de Souza
Non-equilibrium effects in the ^{139}La + ^{40}Ar reaction at 10 MeV per nucleon observed in a study of neutron emission
Phys. Rev., C39, 1845 (1989)

248. W. U. Schröder, J. L. Wile, D. Pade, S. S. Datta, J. Tōke, J. R. Huizenga, and R. T. de Souza
Mass energy flow in damped reactions
Proc. Symp. on Nuclear Physics, Bombay, December 27–31, 1988, Vol. 31A, 231 (1989)

249. W. U. Schröder, J. L. Wile, D. Pade, S. S. Datta, J. Tōke, J. R. Huizenga, and R. T. de Souza
Non-equilibrium energy transport in damped reactions
Proc. Int. Conf. Nuclear Reaction Mechanism, Calcutta, January 3–9, 1989, p. 72, S. Mukherjee, editor (Saha Inst. Nucl. Phys., 1989)

250. J. R. Huizenga, A. N. Bekhami, I. M. Govil, W. U. Schröder, and J. Tōke
Influence of rotation-induced nuclear deformation on α-particle evaporation spectra
Phys. Rev. C40, 668 (1989)

251. J. Tōke, W. U. Schröder, and J. R. Huizenga
Correlations between fragment mass and excitation energy in damped reactions
Phys. Rev. C40, R1577 (1989)

252.   J. R. Huizenga, A. N. Behkami, I. M. Govil, W. U. Schröder, and J. Tōke
       Dependence of the shape of alpha-particle evaporation spectra on nuclear
       deformation
       Proc. Symposium on Nuclear Dynamics and Nuclear Disassembly, 197th National
       ACS Meeting, Dallas, April 9–14, 1989, pp. 228–243, J. B. Natowitz, editor
       (World Scientific Publishing Company, 1989)
253.   J. L. Wile, S. S. Datta, W. U. Schröder, J. R. Huizenga, R. T. de Souza, and
       D. Pade
       Excitation energy equilibrium in damped ^{139}La+^{40}Ar collisions at 15 MeV
       per nucleon
       Phys. Rev. C40, 1700 (1989)
254.   W. U. Schröder and J. R. Huizenga
       Heavy-ion-induced fission—experimental status
       Proc. International Conference "Fifty Years Research in Nuclear Fission," Berlin,
       Germany, April 3–7, 1989, *Nucl. Physics A502,* 473C (1989)
255.   J. L. Wile, S. S. Datta, R. T. de Souza, J. R. Huizenga, D. Pade, W. U.
       Schröder, and J. Tōke
       Evidence for radial-energy scaling of non-equilibrium neutron yields in
       damped ^{139}La $+$ ^{40}Ar reactions
       Phys. Rev. Letters 63, 2551 (1989)
256.   R.W. Atcher, A. M. Friedman, J. R. Huizenga, and R.P. Spencer
       A radionuclide generator for the production of ^{211}Pb and its daughters
       J. Radioanal. Nucl. Chem. Lett. 135, 215 (1989)
257.   J. Tōke, W. U. Schröder, and J. R. Huizenga
       Aspects of kinematical coincidence measurements of excitation energy
       division in damped reactions
       Nuclear Instruments and Methods, A288, 406 (1990)
258.   J. Tōke, R. Planeta, W. U. Schröder, and J. R. Huizenga
       Correlations between energy and mass partition in the damped reaction
       ^{165}Ho $+$ ^{74}Ge at $E_{Lab} = 8.5$ MeV/nucleon
       Phys. Rev. C 44, 390 (1991)
259.   M.B. Chatterjee, S. P. Baldwin, J. R. Huizenga, D. Pade, B. M. Quednau,
       W. U. Schröder, B. M. Szabo, and J. Tōke
       Energy partition in near-barrier strongly damped collisions ^{58}Ni $+$ ^{208}Pb
       Phys. Rev. C 44, R2249 (1991)
260.   J. R. Huizenga
       Cold fusion: the scientific fiasco of the century
       (University of Rochester Press, Rochester, New York, April 1992)
261.   J. R. Huizenga
       Cold fusion labeled "Fiasco of Century"
       FORUM for Applied Research and Public Policy Winter, p. 78 (1992)

262. J. L. Wile, S. S. Datta, W. U. Schröder, J. Tõke, D. Pade, S. P. Baldwin, J. R. Huizenga, B. M. Quednau, R. T. deSouza, and B. M. Szabo
Thermal properties of compound systems formed in fusion of ^{118}Sn and ^{124}Sn nuclei with ^{28}Si
Phys. Rev. C 47, 2135 (1993)

263. J. R. Huizenga
Are the world's energy problems over?
Leaders Magazine 16, no. 1, p. 21 (1993)

264. M.B. Chatterjee, S. P. Baldwin, J. R. Huizenga, D. Pade, B. M. Quednau, W. U. Schröder, B. M. Szabo, and J. Tõke
Reply to "energy partition in near-barrier strongly damped collisions ^{58}Ni + ^{208}Pb"
Phys. Rev. C47, 3003 (1993)

265. J. R. Huizenga
Size of the periodic table
Journal of Chemical Education 70, no. 9, 730 (1993)

266. J. R. Huizenga
Cold fusion: the scientific fiasco of the century
(Oxford University Press, Oxford, New York, 1993)

About the Author

John R. Huizenga is the Tracy H. Harris Professor Emeritus of Chemistry and Physics at the University of Rochester. He is the author of *Cold Fusion: The Scientific Fiasco of the Century* and co-author of *Nuclear Fission* and *Damped Nuclear Reactions*. Among his professional awards are a Fulbright Fellowship (1954–55), two Guggenheim Fellowships (1964–65 and 1973–74), the Department of Energy's E. O. Lawrence Memorial Award (1966), the American Chemical Society's Glenn T. Seaborg Award (1975), Calvin College's Distinguished Alumni Award (1975), and the Leroy Randle Grumman Medal for Outstanding Scientific Achievement (1991). Dr. Huizenga is a member of the National Academy of Sciences and the American Chemical Society, and a Fellow of the American Academy of Arts and Sciences, the American Physical Society, and the American Association for the Advancement of Science.